SPACE SCIENCE IN THE TWENTY-FIRST CENTURY: IMPERATIVES FOR THE DECADES 1995 TO 2015

LIFE SCIENCES

Task Group on Life Sciences
Space Science Board
Commission on Physical Sciences, Mathematics, and Resources
National Research Council

NATIONAL ACADEMY PRESS
Washington, D.C. 1988

National Academy Press • 2101 Constitution Avenue, N.W. • Washington, D. C. 20418

NOTICE: The project that is the subject of this report was approved by the Governing Board of the National Research Council, whose members are drawn from the councils of the National Academy of Sciences, the National Academy of Engineering, and the Institute of Medicine. The members of the committee responsible for the report were chosen for their special competences and with regard for appropriate balance.

This report has been reviewed by a group other than the authors according to procedures approved by a Report Review Committee consisting of members of the National Academy of Sciences, the National Academy of Engineering, and the Institute of Medicine.

The National Academy of Sciences is a private, nonprofit, self-perpetuating society of distinguished scholars engaged in scientific and engineering research, dedicated to the furtherance of science and technology and to their use for the general welfare. Upon the authority of the charter granted to it by the Congress in 1863, the Academy has a mandate that requires it to advise the federal government on scientific and technical matters. Dr. Frank Press is president of the National Academy of Sciences.

The National Academy of Engineering was established in 1964, under the charter of the National Academy of Sciences, as a parallel organization of outstanding engineers. It is autonomous in its administration and in the selection of its members, sharing with the National Academy of Sciences the responsibility for advising the federal government. The National Academy of Engineering also sponsors engineering programs aimed at meeting national needs, encourages education and research, and recognizes the superior achievements of engineers. Dr. Robert M. White is president of the National Academy of Engineering.

The Institute of Medicine was established in 1970 by the National Academy of Sciences to secure the services of eminent members of appropriate professions in the examination of policy matters pertaining to the health of the public. The Institute acts under the responsibility given to the National Academy of Sciences by its congressional charter to be an adviser to the federal government and, upon its own initiative, to identify issues of medical care, research, and education. Dr. Samuel O. Thier is president of the Institute of Medicine.

The National Research Council was organized by the National Academy of Sciences in 1916 to associate the broad community of science and technology with the Academy's purposes of furthering knowledge and advising the federal government. Functioning in accordance with general policies determined by the Academy, the Council has become the principal operating agency of both the National Academy of Sciences and the National Academy of Engineering in providing services to the government, the public, and the scientific and engineering communities. The Council is administered jointly by both Academies and the Institute of Medicine. Dr. Frank Press and Dr. Robert M. White are chairman and vice chairman, respectively, of the National Research Council.

Support for this project was provided by Contract NASW 3482 between the National Academy of Sciences and the National Aeronautics and Space Administration.

Library of Congress Catalog Card Number 87-43334

ISBN 0-309-03880-4

Printed in the United States of America

TASK GROUP ON LIFE SCIENCES

Scott Swisher, Michigan State University, Co-Chairman
David Usher, Cornell University, Co-Chairman
Meinrat Andreae, Florida State University
Stanley Awramik, University of California, Santa Barbara
Robert Berliner, Pew Scholars Program, Yale University
William DeCampli, Stanford Medical Center
James Ferris, Rensselaer Polytechnic Institute
Robert Fowles, University of Utah
Andrew Knoll, Harvard University
Robert Kretsinger, University of Virginia
Lynn Margulis, Boston University
Raymond Murray, Michigan State University
Quentin Myrvik, Wake Forest University
John Oro, University of Houston
Tobias Owen, State University of New York at Stony Brook
Donald D. Trunkey, Oregon Health Services University
G. Donald Whedon, International Shrine Hospital
David White, Florida State University
Richard J. Wurtman, Massachusetts Institute of Technology
Richard Young, RCA Government Services

Ex Officio

Jay M. Goldberg, University of Chicago
Harold Klein, The University of Santa Clara

NASA Liaisons

Sherwood Chang
Lawrence F. Dietlein

Joyce M. Purcell, *Staff Officer*
Judith L. Estep, *Secretary*

STEERING GROUP

Thomas M. Donahue, University of Michigan, Chairman
Don L. Anderson, California Institute of Technology
D. James Baker, Joint Oceanographic Institutions, Inc.
Robert W. Berliner, Pew Scholars Program, Yale University
Bernard F. Burke, Massachusetts Institute of Technology
A. G. W. Cameron, Harvard College Observatory
George B. Field, Center for Astrophysics, Harvard University
Herbert Friedman, Naval Research Laboratory
Donald M. Hunten, University of Arizona
Francis S. Johnson, University of Texas at Dallas
Robert Kretsinger, University of Virginia
Stamatios M. Krimigis, Applied Physics Laboratory
Eugene H. Levy, University of Arizona
Frank B. McDonald, NASA Headquarters
John E. Naugle, Chevy Chase, Maryland
Joseph M. Reynolds, The Louisiana State University
Frederick L. Scarf, TRW Systems Park
Scott N. Swisher, Michigan State University
David A. Usher, Cornell University
James A. Van Allen, University of Iowa
Rainer Weiss, Massachusetts Institute of Technology

Dean P. Kastel, *Study Director*
Ceres M. Rangos, *Secretary*

SPACE SCIENCE BOARD

Thomas M. Donahue, University of Michigan, Chairman
Philip H. Abelson, American Association for the Advancement of Science
Roger D. Blandford, California Institute of Technology
Larry W. Esposito, University of Colorado
Jonathan E. Grindlay, Center for Astrophysics
Donald N. B. Hall, University of Hawaii
Andrew P. Ingersoll, California Institute of Technology
William M. Kaula, NOAA
Harold P. Klein, The University of Santa Clara
John W. Leibacher, National Solar Observatory
Michael Mendillo, Boston University
Robert O. Pepin, University of Minnesota
Roger J. Phillips, Southern Methodist University
David M. Raup, University of Chicago
Christopher T. Russell, University of California, Los Angeles
Blair D. Savage, University of Wisconsin
John A. Simpson, Enrico Fermi Institute, University of Chicago
George L. Siscoe, University of California, Los Angeles
L. Dennis Smith, Purdue University
Darrell F. Strobel, Johns Hopkins University
Byron D. Tapley, University of Texas at Austin

Dean P. Kastel, *Staff Director*
Ceres M. Rangos, *Secretary*

COMMISSION ON PHYSICAL SCIENCES, MATHEMATICS, AND RESOURCES

Norman Hackerman, Robert A. Welch Foundation, Chairman
George F. Carrier, Harvard University
Dean E. Eastman, IBM Corporation
Marye Anne Fox, University of Texas
Gerhart Friedlander, Brookhaven National Laboratory
Lawrence W. Funkhouser, Chevron Corporation (retired)
Phillip A. Griffiths, Duke University
J. Ross Macdonald, University of North Carolina, Chapel Hill
Charles J. Mankin, Oklahoma Geological Survey
Perry L. McCarty, Stanford University
Jack E. Oliver, Cornell University
Jeremiah P. Ostriker, Princeton University Observatory
William D. Phillips, Mallinckrodt, Inc.
Denis J. Prager, MacArthur Foundation
David M. Raup, University of Chicago
Richard J. Reed, University of Washington
Robert E. Sievers, University of Colorado
Larry L. Smarr, National Center for Supercomputing Applications
Edward C. Stone, Jr., California Institute of Technology
Karl K. Turekian, Yale University
George W. Wetherill, Carnegie Institution of Washington
Irving Wladawsky-Berger, IBM Corporation

Raphael G. Kasper, *Executive Director*
Lawrence E. McCray, *Associate Executive Director*

Foreword

Early in 1984, NASA asked the Space Science Board to undertake a study to determine the principal scientific issues that the disciplines of space science would face during the period from about 1995 to 2015. This request was made partly because NASA expected the Space Station to become available at the beginning of this period, and partly because the missions needed to implement research strategies previously developed by the various committees of the board should have been launched or their development under way by that time. A two-year study was called for. To carry out the study the board put together task groups in earth sciences, planetary and lunar exploration, solar system space physics, astronomy and astrophysics, fundamental physics and chemistry (relativistic gravitation and microgravity sciences), and life sciences. Responsibility for the study was vested in a steering group whose members consisted of task group chairmen plus other senior representatives of the space science disciplines. To the board's good fortune, distinguished scientists from many countries other than the United States participated in this study.

The findings of the study are published in seven volumes: six task group reports, of which this volume is one, and an overview report of the steering group. I commend this and all the task group reports to the reader for an understanding of the challenges that confront the space sciences and the insights they promise for the next century. The official recommendations of the study are those to be found in the steering group overview.

> Thomas M. Donahue, Chairman
> Space Science Board

Contents

1. INTRODUCTION   1
   Overview, 1
   Exobiology, 2
   Global Biology/Biospheric Science, 3
   Controlled Ecological Life Support System
   (CELSS), 4
   Space Biology, 5
   Human Biology and Space Medicine, 6
   Implementation, 7

2. EXOBIOLOGY   8
   What is Exobiology?, 8
   Planetary Exploration and the Need for Space Data, 12
   Research Topics, 15
   Space Missions, 39
   Conclusions and Recommendations, 57

3. GLOBAL BIOLOGY/BIOSPHERIC SCIENCE   59
   Background, 59
   Biosphere-Atmosphere Interactions, 61
   Global Ecology, 65

4. CONTROLLED ECOLOGICAL LIFE SUPPORT
   SYSTEM (CELSS)   69
   Definition, 69
   Research Objectives, 70
   Accomplishments, 71

5. SPACE BIOLOGY   72
   The Problems, 72
   Work To Date, 74
   Future Work, 75

6. HUMAN BIOLOGY AND SPACE MEDICINE   77
   Introduction, 77
   Experimental Use of Animals, 78
   Neurosensory Physiology, 79
   Bone and Mineral Metabolism, 87
   Muscle Metabolism, 93
   Cardiovascular, Pulmonary, and Renal Systems, 95
   Integrated Functions, 101
   Radiation Effects, 108
   Behavior and Performance, 117
   Health Maintenance, 123

7. INTERNATIONAL COOPERATION IN SPACE LIFE
   SCIENCES   130
   Ongoing Agreements and Memoranda of Understanding, 131
   Agreements Pending, 132
   Agreements Finished, 132

8. INSTRUMENTATION AND TECHNOLOGY   134
   Introduction, 134
   Exobiology, 135
   Global Biology, 136
   Space Biology, 136
   Space Medicine, 138
   Controlled Ecological Life Support System
   (CELSS), 140
   Computation, Integration, and Robotics, 142

1
Introduction

OVERVIEW

The life sciences have been and will continue to be an integral part of our space program. As anticipated in the legend of Icarus, flight can expose us to anoxia and extremes of temperature, and spaceflight adds microgravity and radiation. We cannot adapt to these conditions or protect ourselves from their effects without a sophisticated understanding of the underlying physiological responses. On spaceflights lasting for months, recycling wastes becomes economically attractive; on flights lasting for years, controlled ecological life support systems are imperative. Research into several fundamental problems in biology—plant growth, biomineralization, vestibular function, and development—will also benefit from access to microgravity laboratories.

We are seeing the birth of a new science that combines the global perspective of the earth sciences with the principles of ecology. NASA has the expertise and the organization to be a major contributor to a global study of the interactions of the biota with the atmosphere, hydrosphere, and geosphere. A greater understanding of our biosphere will have a profound impact on our international relations and on our economy.

Speculations on how life began have occupied some of the

best minds for millennia. The task group believes this is now a soluble scientific problem. NASA can take a lead in the integration of planetary sciences, molecular biology, and prebiotic chemistry. The result will be a new understanding of our own origin and evolution, and a more reliable estimate of the possibility of life outside our solar system. The intellectual impact of exobiology and global biology will probably equal that of molecular biology.

The four disciplines treated in this report—exobiology, global biology, space biology, and space medicine—span an extremely broad range of intellectual subject matter and technology. Their parent disciplines—ecology, molecular biology, chemistry, astronomy, and medicine—are well established. But there have been so few flight opportunities for studies in these fields, especially space biology and medicine, that they have yet to develop into mature space sciences. They are all young disciplines, still defining their basic questions and strategies. They are united by the study of life, and especially its evolution.

EXOBIOLOGY

Understanding the origin, early evolution, and distribution of life is the focus of a major scientific effort in NASA. The early environment of the Earth is being deciphered through the study of biological and chemical fossils in 3- to 4-billion-year-old rocks. Within our own solar system there are strong indications of organic reactions on the surfaces and in the atmospheres of several planets, on the satellite Titan, and in comets and asteroids. Organic molecules, many of which are constituents of living organisms, have been detected in meteorites as well as in interstellar space. Exciting discoveries of molecules synthesized in the laboratory under conditions presumed to exist on the primitive Earth have led to theoretical pathways concerning the origin of life on Earth.

The current view, which is gradually being confirmed, is that the chemicals of life abound in the universe and that the conditions that gave rise to life on Earth may exist in other places. We do not yet know the details of this chemistry, nor do we know whether life has actually arisen elsewhere in our own solar system or beyond. We do not even know for certain whether planets exist outside of our own solar system, although there is good reason to believe they do. This is an area of research that is tractable to both laboratory experimentation and space exploration and involves a broad range of interdisciplinary collaborations.

By 1995, the task group expects that our capabilities in the field of exobiology will have expanded markedly. By then we should be able to probe comets directly through chemical analysis and to identify and quantify the organic molecules in these bodies and their relationship to early planetary history. We should also be in a position to determine with some confidence the presence (or absence) of other planetary systems. We should be able to collect cosmic dust particles in space for detailed physical and chemical analysis, particularly for organic content. At the same time, we should be able to extend the search for clues to the history of life to other planets, particularly Mars, where only preliminary studies were done by Viking, and to Titan, Europa, and perhaps other satellites of other planets. Studies of these bodies should include a search for evidence of life forms that once existed, but are no longer present. By 1995, we should also have the ability to search our galaxy by means of radiotelescopes for signs of intelligent civilizations.

GLOBAL BIOLOGY/BIOSPHERIC SCIENCE

The ability to travel in space has revolutionized our perception of the universe and our place within it. Humans can now view their planet from afar and contemplate its entirety while at the same time applying their scientific armamentarium to an array of problems not approachable by any other means.

Earth is essential to human existence; it is the only planet known to harbor life. Our understanding of the evolutionary relationships between living organisms and the planet is limited and based on local or regional data gathered over the years by ground-based observations. Spacecraft provide the means of obtaining a global perspective, that is, of looking at and measuring key phenomena globally and continuously.

Fundamental to understanding the biosphere is deciphering the interrelationships between biological processes and geochemical-geophysical processes. For example, study of biogeochemical cycles through study of changes in atmospheric carbon dioxide and periodic measurements of global biomass and productivity is now both possible and timely. Monitoring biosphere-climate interactions, measuring biogenic aerosols, monitoring surface changes induced by phenomena such as deforestation, desertification, and agriculture, and measuring oceanic productivity are all activities that can be carried out from space. Interactions between the biosphere and the atmosphere can also be measured from space. These include the exchange of trace gases between the biosphere and atmosphere, the effects of biomass burning, tropospheric chemical cycling, and stratospheric contamination.

Earth-orbiting spacecraft offer the exciting prospect of monitoring environmental conditions relevant to certain disease outbreaks, such as malaria. By global monitoring of important variables such as seasonal rainfall and temperature, predictions of outbreaks of mosquito populations can be made. Such studies will allow much more effective modeling of global ecology. This, in turn, will permit a recognition and understanding of threatening trends.

The surface of the Earth, viewed in terms of temperature, water content, sediments, and atmospheric composition, is completely different from that predicted as intermediate between Venus and Mars. To understand our planet we must understand the cumulative impact of 4 billion years of life.

CONTROLLED ECOLOGICAL LIFE SUPPORT SYSTEM (CELSS)

Human exploration of our solar system will require missions of long duration. These, in turn, require not only our understanding of human tolerance and limitations, they also present extremely complex technical and theoretical problems of providing the air, water, and food for a livable environment. Eventually we will no longer be able to carry from Earth sufficient supplies to support extended space travel. The mere weight and volume of these expendables will be beyond the carrying capacity of the spacecraft. We will be forced to recycle ever more of these materials. Air must be cleaned and humidified or dehumidified, water purified and reused, and food produced, consumed, and the wastes processed and recycled. Virtually nothing can be discarded in the tightly closed systems required for explorations of several years' duration. These systems must be thoroughly evaluated in flight prior to planning long missions.

As formidable as these engineering problems are, the biological problems of such a life support system may prove even more difficult, especially if the human palate and psyche demand the presence of higher plants. The effects of microgravity on plant reproduction, development, and growth—especially when coupled with those of artificial lighting—are not understood. Success in this endeavor demands imaginative cooperation between engineers, chemists, nutritionists, and ecologists. Aside from the utility of a Controlled Ecological Life Support System (CELSS) as a life support aid, the concept of closing, at least partially, an artifical ecosystem is of interest to the science of ecology and may offer a research tool of considerable value for study of the principles by which nature's ecosystems function.

SPACE BIOLOGY

Throughout its evolution, life on Earth has experienced only a one-g environment. The influence of this omnipresent force is not well understood, except that there is clearly a biological response to gravity in the structure and functioning of living organisms. The plant world has evolved gravity sensors; roots grow "down" and shoots "up." Animals have gravity sensors in the inner ear. Many fertilized eggs and developing embryos orient their cleavage planes relative to the gravity vector. Access to a microgravity space station laboratory will facilitate research on the cellular and molecular mechanisms involved in sensing forces as low as 0.001-g and subsequently transducing this signal to a neural or hormonal signal. A major challenge to our understanding and mastery of these biological responses is to propagate selected species of higher plants and mammals through several generations at microgravity.

As was amply demonstrated by Pasteur, as well as countless successors, investigations in medicine and in agriculture contribute to and benefit from basic research. Understanding the responses of humans and of plants to microgravity has enormous practical significance for manned spaceflight. The use of microgravity to eliminate microconvection in crystal growth, in electrophoresis, and in biochemical reactions should continue to be evaluated for both research and commercial application. Conversely, the urgent need to moderate the debilitating effects of bone and muscle wasting may lead to fundamentally new insights into biomineralization and controls of gene transcription and translation. Although serendipity is hardly the basis of a research strategy, we emphasize the value to science in general—and to biology in particular—of creating a research environment in which a creative scientist can observe unanticipated phenomena. These questions then become the stuff of logical analysis and formal reports.

HUMAN BIOLOGY AND SPACE MEDICINE

Our space program should develop the capability for manned space missions of several years. So few data are available that any projection is tentative. The physiological effects of short-duration spaceflight will probably be tolerated or compensated for, if not well understood and solved, by the middle of the Space Station era (approximately 2005). However, the long-term effects of microgravity, or even the reduced gravity of the Moon, on bone and muscle metabolism and on cardiovascular function will probably remain poorly understood.

Crew members are protected from ion radiation by the Earth's magnetic field while in the low-inclination, low-altitude orbits of the Shuttle and the Space Station. However, they would be exposed to significant heavy ion radiation during interplanetary missions or while inhabiting a lunar or martian base. This exposure could have disastrous effects on the central nervous system, because heavy ion radiation has recently been shown to inflict "single hit" damage, even death, on nondividing cells.

The more general problem of the ability of human beings to thrive in a closed, stressful environment assumes novel importance and exigency with extended spaceflights. In addition to the problems of weightlessness and heavy ion radiation, the crew may have to deal with increased microbial density in the cabin air, organic and inorganic toxins (outgassing products), nutritional limitations, and the problems of health care delivery in space. These physical stresses will exacerbate the severe emotional stresses associated with working and living in confined quarters. Many of these problems have no terrestrial analog and must be understood in much greater depth before we can permit a manned mission to Mars.

Some of the research in space biology and medicine is concerned with the health and welfare of the astronauts. Other components are of basic scientific interest and deal with fundamental questions concerning the role of gravity in life processes. The task group believes that these two objectives complement one another.

IMPLEMENTATION

The following chapters discuss the status and goals of these five areas of research—exobiology, global biology, controlled exological life support systems, space biology, and space medicine. Chapter 8 discusses the instrumentation and technologies required to achieve these goals. The task group emphasizes that research on living organisms, including humans, imposes constraints not encountered in the other space sciences. On the other hand, many of the instruments, as well as the strategies, of the global biologists are common to the earth scientists. Similarly, the section treating exobiology contains numerous cross-references to the field of planetary exploration.

2
Exobiology

WHAT IS EXOBIOLOGY?

Throughout history, humanity's creation myths appear to reflect each culture's perception of the dimensions of its universe and its place within it. Today, the scope of those perceptions has expanded beyond the reaches of the solar system to the stars and the interstellar clouds that populate the seemingly limitless expanse of space. We see life as the product of countless changes in the form of primordial stellar matter wrought by the processes of astrophysical, planetary, and biological evolution. The science of exobiology attempts to reconstruct the natural history of processes and events involved in the transformations of the biogenic elements from their origins in nucleosyntheses to their participation in Darwinian evolution in the solar system on planet Earth. From this reconstruction will emerge a general theory for the evolution of living systems from inanimate matter.

The goal of exobiology is to increase knowledge of the origin, evolution, and distribution of life in the universe. This is a multidisciplinary science, and the conceptual and experimental tools of virtually all scientific disciplines and branches of learning are relevant. In seeking answers to such questions as how the development of the solar system and its planets led to the origin of life on Earth, how planetary evolution subsequently influenced the course of biological evolution, and where else life may be found in the solar system and beyond, exobiology brings together life scientists and physical scientists in a common interest.

Exobiology is concerned with four evolutionary epochs: (1) cosmic evolution of biogenic elements and compounds; (2) prebiotic evolution; (3) early evolution of life; and (4) evolution of advanced life. Each of these epochs, briefly described below, represents a major arena of research. Between now and the mid-1990s, the task group expects this conceptual framework to become widely acknowledged and to act as a stimulus for interdisciplinary attacks on exobiological research problems.

1. *Cosmic Evolution of Biogenic Elements and Compounds.* The first epoch encompasses galactic time and distance scales and involves the death and birth of stars. It begins with the synthesis in stars of the biogenic elements—the elements that make up all life—and their ejection into the interstellar medium; it ends with the distribution of these elements and their compounds throughout our solar system within the planetoids, which became building blocks of planets. Discoveries in carbonaceous meteorites strengthen this perspective, as organic and mineral matter made up of carbon, nitrogen, hydrogen, and oxygen has been found that retains properties traceable to its origins in interstellar clouds and stars.

How commonly the aggregation of interstellar dust and gas into small primitive bodies occurs during star formation is not known. What transformations were undergone by the biogenic elements and their compounds during this process remain poorly understood, as are the ways in which the physical and chemical properties of these elements and their compounds may have influenced the course of events during the formation of the solar system.

Answers to these questions will develop, however, as astrophysicists and astrochemists take advantage of the capabilities for large-scale modeling, and make use of sensitive space-borne astronomical telescopes with high spatial and spectral resolution to make observations of condensed matter in protostellar regions. In ground-based laboratories and by remote spacecraft, studies of interstellar dust and samples of other relict material—meteorites, comets, asteroids, and interplanetary dust—will continue to help reconstruct the nature and chronology of the processes that took place at the time of the solar system's formation.

2. *Prebiotic Evolution.* This epoch begins with the accretion of planets and ends, in the case of Earth, with the emergence of living organisms after nearly 1 billion years. This is the epoch for which there has been found no geological record on Earth and, therefore, no direct basis on which to reconstruct the conditions of the environment. Yet it is the period in which life emerged from inanimate matter. For any planet this is the period of most rapid environmental change as the energy of accretion and radionuclide decay dissipates, and the planet undergoes a transition to either geological inactivity or to a steady state. Prebiotic chemical evolution is inseparable from planetary evolution, and the path that leads to the origin of life may be terminated if the development of a planet takes the wrong turn.

In what ways and how fast were conditions changing on Earth in its first billion years when life originated? Were conditions initially similar on Mars or Venus? What types of geological settings were maintained far from equilibrium and conducive to the origin of life? What processes and reactions involving the biogenic elements were important for the origin of life?

Answers to these questions can come from several sources: the application of planetary geophysics and geochemistry to the development of models for Earth's earliest history, the deciphering of the existing geological record among the terrestrial planets, and laboratory simulations of prebiotic reactions.

3. *Early Evolution of Life.* The third epoch begins with the emergence of living systems prior to 3.5 billion years ago, and carries through to the evolution of multicellular organisms, which appear in the fossil record about 1 billion years ago. The history of life in this period is largely the history of microscopic unicellular organisms. This period is also a time of continued change in environmental conditions, although less intensive or rapid than in the previous epoch, as the physical evolution of Earth moved into a more moderate stage. The Sun gradually became more luminous, the frequency of large-scale accretionary impacts declined, the Moon's separation from Earth increased, volcanism declined, the continents grew in volume and wandered across the face of the planet, oxygen began to appear in the atmosphere, and the geomagnetic poles reversed.

In this epoch the major questions concern the relationship between the evolution of unicellular life and the climate and changing geology of the Earth. What was the path by which the ancestors of modern microbes evolved from the first living organisms? In what ways did physical and chemical changes in the environment influence the rate and direction of microbial evolution? Over what geological time scales did major events in microbial evolution occur? How did the evolving biota modify and modulate their environment over time? What was the nature of the earliest forms of life, and in what sequence were new attributes acquired? What are the simplest biochemical mechanisms and biophysical structures that can fulfill the functions of living systems, and what irreducible combinations of these constitute a living entity? Did life arise on Mars or on other planets during this time, and if so what changes in those extraterrestrial environments would have led to its extinction?

The answers to these questions hinge on our ability to identify and obtain inorganic and organic fossils, and to decipher the record of geological and biological evolution through the layers of alteration and mutation accumulated over several billion years. This means that the right rocks and organisms must be studied; the phylogeny of microbial life must be tied to the chronology of geological change; and the time resolution with which we can discern changes in these two records must become more finely tuned.

4. *Evolution of Advanced Life.* The fourth epoch deals with the most recent billion years of the history of life, in which multicellular plants and animals and intelligent species evolved. Exobiological interest in this epoch grew out of the realization that collisions of asteroidal-sized objects with Earth produce global changes in its surface environment. That these changes would perturb the biosphere cannot be denied; whether they can account for major extinctions in the course of biological evolution, particularly events that may have placed advanced life on its evolutionary track to intelligence, remains to be determined. What is clear, however, is the importance of gaining more knowledge about the relationship between biological evolution and changes in the environment.

PLANETARY EXPLORATION AND THE NEED FOR SPACE DATA

In this search for knowledge about the origin of life as a natural process, NASA must explore as many extraterrestrial bodies as possible for the relevant information they can supply. The search should include bodies totally devoid of organic chemicals, those conceivably undergoing (or having underdone) organic chemical evolution, and those possibly harboring life.

The study of lifeless planets provides examples of environments where chemical or biological evolution ended. On them it may be possible to find the remnants of chemical evolution or of past life and to learn how planetary changes may have broken the thread of chemical or biological evolution. The discovery that there is no life, extant or extinct, or no organic matter on a planet is of high interest because the conditions on the planet and what we can learn of its past history constitute basic data pertinent to a general theory of the origin of life.

Mars continues to be an extremely interesting site for exploration and study for exobiology. Although the Viking mission found no evidence of extant life, the search was limited and not directed at optimum sites. In recent years, dicoveries about the earliest history of life on Earth have reemphasized the need to examine the geological record of Mars. Compelling evidence for flourishing communities of microorganisms has been found in 3.5-billion-year-old sedimentary rocks of Australia and South Africa. Even if no martian life exists, the evidence of liquid water and a more clement epoch in the first billion years of martian history, at the time when Earth already had a thriving microbial biosphere, has important implications. The possibility that life arose on Mars early on and subsequently became extinct must be kept open and investigated whenever the opportunity to send missions to Mars arises. In addition, Mars exploration may provide a geological record of the first billion years of planetary evolution, for which little trace has yet been found on Earth.

The Mars Orbiter Mission (MOM) will be invaluable for providing data that will aid in identifying sedimentary and other types of environments that have resulted from the interaction of liquid water with the planetary surface. Further characterization of such sites should be carried out with landers or penetrators capable of making in situ measurements both at the surface and at depth. These measurements should provide further analysis of the chemistry and mineralogy of the surface and subsurface rocks. They should determine the nature and abundances of the forms in which the biogenic elements occur. These explorations should provide a sound basis for site selection for future sample return missions.

The unusual chemistry of the martian surface soils, manifested in the biomimetic responses elicited by the Viking biology experiments, continues to be intriguing and inadequately understood; it may contain important clues to the role of minerals and inorganic chemistry in prebiotic evolution. Elucidation of this chemistry may be attained in part by conducting experiments with landers and penetrators, but a full understanding will probably require detailed chemical, mineralogical, and geochronological study in earth laboratories of a sample returned from Mars.

It is conceivable that a sample return mission could be mounted early in the first decade of the twenty-first century, in which case it should be given highest mission priority for exobiology. By the mid-1990s, planning for integrated exobiological and geological field investigations should get under way in the event that sample return missions will involve human rather than robotic activity on the martian surface. Microscale methods of analysis for characterizing and determining the origin of chemical and mineral phases composed of the biogenic elements should be developed for use in both remote and terrestrial laboratories.

Comet sample return is high on the mission priority list. Earth-based astronomical observations of comets, together with the Halley encounters, have provided a wealth of data on the composition of volatile gases in the coma and the properties of the dust. From these observations conclusions may be drawn about what comets are composed of and how they behave while under the influence of the Sun's radiation, gravitation, and magnetic fields. But major questions about comets remain unanswered—their origin, the identity of higher molecular weight organic compounds, the nature of material contributed by interstellar and solar nebular sources, the ages of their components, and their histories of accretion and thermal and dynamical evolution.

All of these questions are pertinent to exobiology, insofar as comets are composed largely of water, organic matter, and other materials containing the biogenic elements. Comets provide a "fossil" record of the materials and processes involved in the transition from the diffuse realm of gas and dust represented by interstellar clouds to the highly condensed realm of planetary objects. As with carbonaceous meteorites, which are thought to have been derived from either primitive asteroids or "burnt out" comets (or both), this record can best be studied with samples. The maximum scientific return will be obtained from samples that have been collected and preserved in a state as close as possible to that of their original storage in a comet.

Comets and asteroids are the remnants of star formation in a system in which planets formed and life arose. Detection of planets orbiting other stars would represent a major milestone in establishing the generality of planetary solar systems and increasing the probability that life could have arisen elsewhere in the galaxy. Searches for cometary and asteroidal matter associated with other stars or with protostellar objects will provide a basis for determining the frequency of occurrence of preplanetary matter containing the biogenic elements. By the turn of the century, capable astrometric and infrared telescopes should be available, and searches for planetary and preplanetary objects should have broadened. In the outer solar system, Titan will continue to be a target of interest for exobiology; it represents a natural laboratory for the study of planetary organic chemistry. Titan provides a Urey-Miller (in contrast to a Rubey) type model for Earth's prebiotic atmosphere, one which contains nitrogen and hydrogen and in which methane is the predominant carbon source. The variety of minor atmospheric constituents identified by the Voyager missions and earth-based astronomical observations already testifies to the organic chemistry taking place in Titan's atmosphere. However, the origin of the major gases in the atmosphere, their relationship to the accretionary and outgassing history of the satellite, the full range of complexity in the organic chemistry occurring in the atmosphere, and what process is responsible for which observed minor compounds will remain poorly understood. A Titan atmospheric probe and a lander, capable of both characterizing the molecular and isotopic compositions of materials composed of the biogenic elements as well as measuring the abundances and isotopic compositions of the noble gases, would address many of these questions.

Venus, with its atmospheric water, and the Galilean satellites Callisto, Ganymede, and Europa, with their water ice, are objects that contain clues to the history of water in planetary environments. Whether Venus has lost most of its initial aqueous endowment, which may have been comparable in size to that of Earth, or whether it was always strongly depleted in water is an issue pertinent to understanding the mechanisms that distributed and preserved the biogenic elements among the terrestrial planets and perhaps made Earth unique among them. The Galilean satellites, if built up from planetesimals resembling carbonaceous chondrites and comets, may also contain a frozen record of the organic matter that survived planetary accretion. Venus and the Galilean satellites will be pertinent targets of exobiological interest well into the next century.

RESEARCH TOPICS

Formation and Evolution of Biogenic Elements and Compounds

The biogenic elements are those that compose the bulk of life and are generally thought to be essential for all living systems. Primary emphasis is placed on the elements hydrogen, carbon, nitrogen, oxygen, sulfur, and phosphorus. The compounds of major interest include water and those normally associated with organic chemistry. The essential elements usually associated with inorganic rather than organic chemistry are also included—e.g., iron, magnesium, calcium, sodium, potassium, and chlorine—but they are given secondary emphasis. Water plays a central role in the development of life as we know it, and therefore in the environments in which chemical evolution could have occurred. Thus, special importance is attributed to the cosmic history of water and its interaction with other substances of either organic or inorganic nature.

It is useful to identify six stages in the cosmic history of the biogenic elements and compounds: (1) nucleosynthesis and ejection to the interstellar medium, (2) chemical evolution in the interstellar medium, (3) protostellar collapse, (4) chemical evolution in the solar nebula, (5) growth of planetesimals from dust, and (6) accumulation and thermal processing of planetoids. In each of these stages there are major scientific questions to be answered.

Nucleosynthesis and Ejection to the Interstellar Medium

This stage begins the cosmic evolution of the biogenic elements. Not only is it responsible for the origin of the elements, it also initiates the condensation of solid matter out of the gas phase.

Astronomical observations should be made of supernovae, novae, late type giant stars, circumstellar shells, and planetary nebulae where biogenic elements are being produced and ejected into the interstellar medium. Whereas emphasis in the past has been placed on gases, future studies should focus on characterizing the dust and grains and determining the extent to which these primordial solid condensates survive the transit from sites of stellar origin to interstellar clouds. Especially useful will be high-resolution measurements at millimeter and infrared wavelengths. These measurements will require the Large Deployable Reflector (LDR) and the Space Infrared Telescope Facility (SIRTF), which should be in operation by the turn of the century.

Observations of planetary nebulae and the circumstellar shells of late type carbon stars indicate the growth of carbonaceous grains from the gas phase. The mechanisms by which the condensation processes occur are unknown and require elucidation. Although future theoretical simulations of these processes will be undertaken, the Space Station should offer microgravity conditions highly suitable for experimental investigations. Provisions for microparticle research should be included in the Space Station's capabilities.

Chemical Evolution in the Interstellar Medium

Interstellar clouds serve both as the collectors of atomic and dusty debris from stars in terminal stages of evolution and as the spawning grounds of new stars. In the course of cosmic evolution they provide the first environments in which gas-gas and gas-solid interactions occur between water, organic, and inorganic compounds.

By the mid-1990s, the gas phase chemistry of low-molecular-weight compounds in interstellar clouds should be reasonably well known. For lack of observational tools, however, the chemistry and the role of grains in interstellar processes will still be poorly understood. The telescope facilities mentioned in the preceding section should be used to characterize the biogenic elemental composition and organic chemical content of interstellar dust and, thereby, to begin closing this large gap in knowledge. These same facilities should also be employed to search for interstellar methane and carbon dioxide, key starting materials for organic chemical evolution in the solar system.

Knowledge of the properties of high-molecular-weight organic compounds in the gas phase will continue to be important in the next century. This information is necessary to determine the complexity of interstellar chemistry and to allow mechanisms for the formation of these molecules to be determined. With the availability of suitable microwave spectra, sensitive searches for glycine and adenine, which are among the simplest molecular building blocks of proteins and nucleic acids, can be carried out.

Computer modeling of interstellar grain growth and gas-grain interactions should be used to explore the organic chemistry that could occur on or in various types of grains. For experimental approaches to such research, the microgravity conditions attainable on the Space Station could offer advantages over terrestrial environments. The occurrence of molecules predicted on the basis of model studies could be tested by telescopic searches.

The acquisition of intact interstellar grains for detailed laboratory studies of their structure and composition is a very exciting prospect, and should be set as a goal for the turn of the century. Especially important would be application of dating techniques to determine the chronology of the presolar processes that produced the interstellar grains. Throughout this time, substances of interstellar origin should continue to be sought in comets, interstellar dust particles (IDP), and carbonaceous meteorites to establish more firmly the continuity of cosmic evolution from interstellar cloud to solar nebula.

Protostellar Collapse

This stage in cosmic evolution encompasses the transition from interstellar cloud to the nascent solar system. During protostellar collapse, while temperatures remain at or below 20K, the concentration of gas and dust undergoes enormous change over approximately 7 orders of magnitude from the highly diffuse conditions of interstellar clouds to the considerably denser state of the solar nebula.

As more and more regions of protostar formation are observed, biogenic compounds—e.g., CO, CS, H_2CO, and HCN—should come into increasing use as molecular probes to reveal the physical conditions and the variations in chemical composition that occur in these environments. Future observational studies of small-scale structures in dark interstellar clouds should be made with instruments currently unavailable such as a millimeter-wave, very-large-array (VLA) radio telescope, or an orbiting LDR. Further developments in astrophysical theory coupled with these observations are expected to yield self-consistent models of collapse dynamics. These models should be used for assessing the extent to which observed inhomogeneous distributions of the organic compounds and water in interstellar clouds are preserved during collapse. As knowledge of collapse kinetics and conditions grows, calculations of increasing sophistication and reliability should be carried out to determine the accompanying chemical and isotopic fractionations (redistributions) that occur between the gas phase and dust. These calculations could provide theoretical bounds on the amount and distribution of interstellar matter composed of biogenic elements and compounds that become incorporated into solar system dust and which survive homogenization processes in the solar nebula.

Chemical Evolution in the Solar Nebula

The solar nebula corresponds to the terminal state of protostellar collapse from an interstellar cloud. In this stage, temperatures increase, gas-solid interactions occur readily, energy fluxes increase, turbulent mass transport of matter between environments that differ in temperature and composition can occur, and solid objects larger than interstellar grains begin to accumulate.

Observations and theoretical understanding of protostellar systems in the mid-1990s should yield more tightly constrained models of the solar nebula. Computer models should be developed to simulate the processes that may have contributed to the chemistry and governed the distribution of biogenic elements and compounds within the nebula over time. Among the processes yielding organic compounds and carbonaceous grains that should be studied are photochemistry due to starlight and the early Sun, large-scale electric discharges, ion-molecule reactions in a partially ionized nebula, and reactions of gases on grains.

Confirmation of the relevance of putative nebular processes for cosmic evolution should be sought by direct astronomical observation of species with predicted molecular and isotopic compositions in protostellar systems. In addition, identification of predicted compounds and condensed phases uniquely attributable to nebular sources should be sought in comets, asteroids, and carbonaceous chondrites, all of which are composed in part of components that originated in the nebula.

The acquisition of samples from a cometary nucleus for examination in terrestrial laboratories is a major goal (see the section on space missions, below). Cometary material is especially interesting because it holds the greatest promise of containing primordial material from the earliest history of the solar system.

Growth of Planetesimals from Dust

In this stage of cosmic evolution in the solar nebula, aggregation and accretion of dust occur as particles settle to the mid-plane of the nebula, where they further agglomerate into kilometer-sized bodies. For this stage as well as the preceding two, self-consistent models should have been developed by the mid-1990s that describe the evolution of temperature and other state variables as a function of radial and vertical position over time in the nebula. These models should allow the calculation of the rates of destruction and the rates of growth and coagulation of particles composed of the biogenic elements and compounds, particularly carbonaceous and icy grains.

Accumulation and Thermal Processing of Planetoids

In our own solar system, small bodies were assembled and distributed around the solar system during this stage and their contents altered to varying degrees by the accretion process itself. Other energetic events, the nature and origin of which remain unclear, further influenced their thermal histories. If the parent bodies of meteorites are any indication of the range of possibilities, these histories ranged from very mild, as in the case of carbonaceous meteorites, to extreme, as in the case of differentiated meteorites.

By the 1990s, the study of meteoritic samples in terrestrial laboratories should have identified those components that were preserved intact since accretion, those that were subsequently altered, and those that were produced as a consequence of thermal processing. The emphasis should then shift to determining what processes could have caused thermal alteration of small bodies in the early solar system, and how widespread these processes were.

It is known from the study of carbonaceous meteorites that reactions between water and minerals in at least one parent body yielded clays, carbonates, and other products usually associated with weathering; in these same objects organic compounds and carbonaceous phases occur in abundance. To determine how widespread such processes and products were, mapping the nature and present distribution of the biogenic elements and compounds in the solar system inward of the giant planets should be completed using the spectral reflectance or other spectroscopic properties of the asteroids. The resulting data will help us to understand how these ingredients were distributed in the solar system as a consequence of the dynamics of nebular evolution and planetary accretion. The particular distribution of these elements may have been important for the origin of life on Earth.

Prebiotic and Chemical Evolution

Knowledge of the conditions on the prebiotic Earth can provide bounds on the nature of the environments and the range of chemical systems and mechanistic pathways that can be fruitfully explored as appropriate models for the development of life.

The history of environments and conditions in which life evolved is preserved, albeit incompletely, in a geological record that extends 3.5 billion years back in time. In sharp contrast, the corresponding record for chemical evolution in the first billion years of Earth history is virtually nonexistent. As a consequence, the conditions on the prebiotic Earth must be inferred by extrapolation either backward in time from the existing geological record or forward in time from the initial conditions associated with planetary accretion. Either approach is fraught with uncertainties.

The 3.8-billion-year-old Isua metasediments of Greenland testify to the existence of bodies of liquid water, carbon dioxide in the atmosphere, volcanism, higher heat flow, relatively stable differentiated crust, and the emergence of continental shelf environments, from which the occurrence of weathering, a hydrologic cycle, and a carbon geochemical cycle can be inferred. The 3.5-billion-year-old rocks from South Africa and western Australia record a marine environment dominated by volcanic islands. The microfossil-bearing black cherts in these formations were apparently deposited during relatively quiescent periods between cycles of volcanic eruptions. Pervasive volcanism must have injected dissolved minerals as well as gases into the ancient seas. The early atmosphere must have had little oxygen, as evidenced by the apparent ubiquity of ferrous iron in the surface environment. Beyond these general characteristics, the record of environmental conditions is mute.

The lack of information about environmental conditions is intimidating. The prevailing temperature of the oceans is not known, although the range of local variations could have run from 4°C to the highs of deep-sea hydrothermal vents as they do today. If, as solar evolution theory asserts, the Sun was 70 to 80 percent as luminous then as it is now, the atmosphere must have been better able to retain incoming solar radiant energy or else the ocean would have frozen. Recent theoretical models indicate that the early atmosphere could have contained much carbon dioxide, and that this could have provided the necessary greenhouse effect; however, the actual solution to the problem of the dim Sun, and the atmospheric structure, composition, and pressure 3.8 billion years ago, remains uncertain. The extent and volume of the oceans at that time are unknown, as are the intensity of tidal activity, the day length, and the amount and spectral distribution of sunlight reaching Earth's surface. Details of the chemistry of seawater and the chemistry and mineralogy of sediments await improvements in the ability to see through the overlay of metamorphic and diagenic alteration. The extent to which the geological settings preserved in the rock record are statistically representative of their actual frequency of occurrence is still poorly understood. In general, detailed reconstruction of environmental conditions from the earliest existing record remains a monumental task that is as important for the biological sciences as it is for the earth and atmospheric sciences. Although progress will be made in some areas in the next 10 years, the overall task will remain a challenge for decades to come.

Even if a satisfactory reconstruction could be achieved, it is possible that the environmental conditions at the time of the origin of life differed substantially from those of 3.8 billion years ago. The lack of a record or the inability to extrapolate far enough back in time from the existing record suggests that we use the earth and planetary sciences to construct a theoretical model for the evolution of Earth's prebiotic environment instead. Ideally, the theoretical approach should yield a model that, when extrapolated forward, converges with that inferred by extrapolation of the geological record backward in time. The outcome of the theoretical approach, however, depends on assumptions for the initial boundary conditions for planet formation, most of which are not known with certainty. If this method is to be useful, we will need more accurate estimates of the rate of planetary accretion, the timing and mechanism of core formation, and the nature of early mantle convection, all of which govern the thermal evolution of the planet. This may allow calculation of the time of origin and the composition of the early atmosphere, oceans, and crust. The third approach that can be used to bridge the gap in the geological record is to examine the martian geologic record, which may preserve evidence from this period. Although the martian record will speak most effectively for Mars, comparative planetology may allow extrapolation to the history of Earth. (See also the section on space missions.)

Conditions far from equilibrium certainly occurred throughout the history of Earth, as they do today, at solid-liquid-gas phase boundary regions. These include fumarolic and volcanic vents on continents and continental shelves, deep-sea plate spreading centers, submarine and island-arc volcanic vents, the land surface, and the sea surface. It seems likely that among these environments were the spawning grounds for the first living systems. The exobiological objectives of research in early planetary evolution should be fourfold:

1. To develop models of specific geophysically active boundary regions in which chemical evolution could have occurred during the prebiotic epoch.

2. To provide limits for the range of variations in temperature, pressure, nature, and intensities of fluxes of energy and matter, and the chemical and mineralogical compositions of the boundary regions, all as a function of time.

3. To assess the role of biogenic elements in influencing specific geophysical and geochemical processes that established, maintained, and altered physical-chemical conditions in these regions over time.

4. To determine the occurrence and history of disequilibrium processes involving the biogenic elements and compounds on other planets in the solar system and beyond.

Energy Harvesting, Storage, and Transduction

An energy gradient was essential for the origin of life on the primitive Earth. The Sun was the principal source of this energy, as it is on the Earth today. This energy was manifested directly in the form of solar radiation and indirectly in the form of lightning and thermal energy. Other energy sources included the shock waves resulting from the impact of comets and meteorites in the atmosphere, the heat release resulting from tectonic processes, and heat and radiation resulting from radioactive decay.

It is likely that many of these sources were more powerful when the Earth first formed than they are today. For example, it was discovered recently that young stars have more than a hundredfold stronger flux in the short wavelength region than do stars the age of our Sun. The intense ultraviolet flux would have been effective in photolyzing the small organic and inorganic molecules present in the atmosphere of the primitive Earth. This energy flux would have been intense at the surface of the Earth since there was little or no ozone shield in the primitive atmosphere to absorb this radiation. Energy from shock waves would have been greater because of the high rate of meteoritic and cometary impact on the primitive Earth. This intense meteoritic bombardment would have warmed the Earth's crust as would the heat resulting from the decay of the highly radioactive elements present there.

Some of these energy sources would have been still near their maximum intensity at the time life originated (about 4 billion years ago) and certainly would have fueled the conversion of the simple constituents of the Earth's atmosphere into more complex structures. What is not clear is how the larger organic structures were protected from destruction. For example, the photochemical transformation and destruction of larger organic molecules proceeds at a faster rate, in general, than that of the smaller ones from which they are formed.

Laboratory studies have shown that reactive compounds such as hydrogen cyanide and nitriles are among the reaction products when these energy sources act on mixtures of compounds that simulate the atmosphere of the early Earth. These compounds store energy that can be used to drive other chemical reactions. For example, nitriles spontaneously condense to form polymers, or the energy stored in the nitrile grouping can be used to drive an otherwise energetically unfavorable reaction via coupled chemical processes. Similar reaction coupling has been observed with the pyrophosphate derivatives formed by thermal processes.

In the past 30 years, there have been extensive studies of the conversion of simple organic compounds to more complex molecules using a variety of energy sources. The next decade will see increasing use of computer simulations to predict the chemical effects that energy sources exerted on the atmosphere of the primitive Earth. The capability that the current generation of computers possesses to deal quantitatively with the rates of hundreds of reactions, coupled with the more precise determination of reaction rates, will permit the analysis of complex scenarios for the possible chemical processes that may have taken place on the primitive Earth. Experimental studies will continue that will be designed to answer specific questions concerning the photochemical generation of complex organic molecules. For example, the photochemical transformation of organics absorbed on clays and minerals will be investigated. Little information is available concerning the effects of surface adsorption on the course of photochemical reactions, yet this may have been an important process on the primitive Earth. In addition, the possible use of clays and minerals for the storage and transduction of energy in chemical evolutionary processes will be studied. Computer simulations of the possible atmospheric and crustal processes will continue into the decade of 1995 to 2005.

Early Life

An energy gradient was required to maintain life once it evolved, as it is required to maintain life on Earth today. The energy that primitive life forms used to drive their biochemical machinery may have come from the breakdown of abiotically formed compounds with a high energy content. Alternatively, primitive life processes may have been driven by direct solar radiation. If the first life forms were not autotrophic, it seems likely that this photosynthetic capability evolved rapidly because the oldest known fossils (3.5 billion years old) appear to represent the products and remains of simple photosynthetic organisms.

Primitive heterotrophs would have utilized the same energetic organic compounds, which were essential for the initial formation of living systems. The pyrophosphate bond incorporated into organic or inorganic compounds would likely have been the energy carrier in the breakdown of these compounds. Significantly, energy carriers of this type are still functional in organisms today.

To utilize solar energy, heterotrophs were required to develop a light gathering system. Porphyrins, formed by abiotic processes, may have been incorporated into these primitive life forms for the absorption of long-wavelength solar radiation (visible light). These compounds are reasonable initial light-gathering compounds because they are formed readily from simpler compounds (pyrroles) and absorb light effectively.

Research will be undertaken in the 1985 to 1995 decade on possible abiotic processes leading to pyrroles and porphyrins. In addition, the light-promoted electron transfer properties of porphyrins embedded in clays and lipid membranes will be investigated using sulfide, ferrous ion, and other reducing agents that were likely to have been prevalent on the primitive Earth. In conjunction with these abiotic approaches, investigations will be conducted to elucidate biochemical processes of primitive photosynthetic and heterotrophic organisms still extant.

The next stage of research, to be undertaken in the 1995 to 2015 time period, may be directed toward devising systems that reduce carbon dioxide and carbon monoxide to formaldehyde and simple carbohydrates. Since these gases are likely components of the primitive atmosphere, the first autotrophs must have had the capacity to carry out reductive transformations of this type. The ability to demonstrate such systems in the laboratory will support the postulate that photochemical processes were an important source of energy for primitive forms of life. As mentioned above, knowledge of the photochemical capabilities of primitive organisms still extant will indicate the research areas to be emphasized in the laboratory studies conducted in the 1995 to 2015 time period.

Replication and Transcription

One of the more dramatic advances in the field of chemical evolution in the past 10 years is the template-directed synthesis of RNA in the laboratory. This system involves mixtures of activated ribonucleotides that condense to form an RNA polymer, complementary to the RNA template present in the reaction mixture. With the proper design of both template and monomer, the incorporation of both purine and pyrimidine nucleotides with the complementary RNA can be accomplished.

The polymerization is strongly influenced by small changes in the constituents of the reaction solution. In one example, the use of Pb^{2+} as the catalyst results in the almost exclusive formation of RNA oligomers with bonding that is not present in living systems, while the substitution of Zn^{2+} for Pb^{2+} gives the 3′,5′-linked RNA found in contemporary life. The fidelity attained with Zn^{2+} is close to that observed with some RNA polymerase enzymes.

The formation of the "natural" 3′,5′-linked oligomers was observed also when the 2-methylimidazolide was substituted for 2-imidazolide derivative of the nucleotide. Metal ion catalysis was not required for the oligomerization of these 2-methylimidazolides at 0°C.

The effect of other reagents on this RNA oligomerization reaction is currently under investigation. Possible effects of the addition of polypeptides to the system or the incorporation of amino acids and peptides into the RNA monomers are also being studied.

It has also been demonstrated that DNA templates can be used in place of RNA templates in this RNA synthesis. In one case that has been reported, RNA oligomers twice as long as the template were among the products of the DNA-directed synthesis of RNA. These findings suggest that "sliding" along the template may provide a route from smaller oligomeric templates to large polymeric products.

Over the past 5 years, it has been discovered that RNA can have catalytic properties. Contrary to accepted dogma, the RNA of some RNA-containing proteins was observed to be the site of catalysis, and it was shown that the RNA alone could carry out the reaction normally carried out by the combined RNA and protein subunits. The role of the protein in these RNA-protein complexes is apparently to facilitate the binding of the RNA of the enzyme to the substrate being cleaved, as well as to maintain the RNA in a conformation that possesses optimal catalytic properties. This discovery suggests the exciting possibility that RNA may have served as the site of both catalysis and information storage in the first life forms and that DNA and protein evolved later.

Template-directed RNA polymerization will continue to be

refined in the next 10 years. By 1995, we should have determined whether simple peptides can exert the same control on the specificity of the reaction as has been observed for metal ions or substituents on the imidazolide. Other activated nucleotides, which are more likely to have been present on the primitive Earth, should also have been synthesized. In addition, systems may be developed by 1995 that can use templates containing equal amounts of purines and pyrimidines. Methods should also have been devised to separate the complementary RNA oligomer from its template so that RNA synthesis is not terminated by this tight association.

At the same time we should have a much greater insight into the mechanism by which RNA catalyzes the cleavage of another RNA molecule. With this information, studies will begin on the RNA-catalyzed cleavage and formation of polypeptides, since there is no reason to believe that such processes will not be feasible.

An efficient system for the formation of RNA and DNA from the appropriate monomers and templates should be in place by 1995. The factors that govern the efficiency and specificity of the synthesis should be well understood. Research on the integration of RNA and polypeptide synthesis can then begin as a start toward understanding translation and the origin of the genetic code. At this point, the understanding of RNA catalysis will have proceeded to the stage where it may be possible to design RNA oligomers with well-defined catalytic properties. It should then be possible to ascertain the minimal structural requirements for RNA molecules that may have exhibited both information storage and catalytic properties in the first forms of life.

Translation

In the contemporary cell, the synthesis of a single protein molecule requires the participation of many agents—genetic DNA, messenger RNA, ribosomal RNA, ribosomal proteins, amino acids, transfer RNA, GTP, ATP, and various enzymes—together with initiation, elongation, and termination factors. The overall reaction results in the sequence of bases in a "coding" region of DNA giving rise to a specific sequence of amino acids in the protein, through the intermediary formation of a molecule of messenger RNA. This translation process is complex, and we may reasonably assume that it evolved from a simpler system. Research into the nature of this simpler system is currently being intensified, for the origin of translation represents one of the most significant single steps along the pathway that led to life as we know it. A simple system that carries out translation should show not only recursive formation of the peptide bond, but also relate the sequence of nucleosides along a molecule of RNA to the specific sequence of amino acids in the polypeptide produced. Such a system has not yet been identified, although this subject has captured the imagination of scientists in many fields, and the literature contains many theoretical suggestions. Weak, but quite specific interactions have been found between individual amino acids and combinations of their anticodon nucleotides. In spite of the advances in modern methods for laboratory synthesis of the peptide bond, prebiotic versions of this reaction are not yet known, except for cases that seem unlikely to have been part of a coded system. The plausibility of a system depends both on the type of chemical compounds required and on the complexity of the system. The less complex the system, the more likely it is to have arisen spontaneously over geological time.

It is likely that by 1995 more examples of specific interactions of amino acids and nucleosides will have been reported, although these are likely to be "static" or equilibrium interactions. It is also likely that an efficient peptide bond formation reaction will be known. Based on the rate of progress in the past 15 years, however, it is probable that a working model of a prebiotic translation scheme will not exist until sometime between 1995 and 2015. Such a discovery would have a profound effect on the field of exobiology, and on our view of our own origin.

Clays and Other Minerals as Catalysts

It is likely that minerals had an important role as catalysts for the formation of biological molecules on the early Earth, since we know of many cases where minerals adsorb and catalyze the reaction of organic compounds.

Current research on mineral catalysis is focused on clays. Clays are formed by the weathering of igneous rock and are found abundantly on the Earth. They have also been detected in the 3.8-billion-year-old Isua rock formation as well as in some meteorites, suggesting that clays were also prevalent on the early Earth. In addition, the chemical transformations observed in the Viking mission suggest clays are prevalent on Mars.

Clays consist of aggregates of platelets of aluminosilicates that contain Mg^{2+}, Fe^{2+}, and Fe^{3+} as occasional substitutions for the aluminum. This substitution can result in a net negative charge on the aluminosilicate lattice, which is neutralized by exchangeable cations. In the contemporary environment these cations consist mainly of Na^+, K^+, and Ca^{2+}. There are some examples of significant quantities of transition metals serving as the exchangeable cations on the clays. The concentration of dissolved transition metal ions was probably much higher in the environment of the primitive Earth, so they are likely to have been associated with clays in greater amounts at that time.

The adsorption of organic compounds causes the clay platelets to separate a few angstroms to accommodate small molecules, and 10 Å or more to bind high-molecular-weight polymers. The extent of adsorption is governed by the charge on the organic molecules, the charge density on the clay surface, the exchangeable cations associated with the clay, and the ionic strength of the medium. Adsorption may result in the hydrolysis, oxidation, reduction, or polymerization of the organic molecule due to the presence of activated water molecules, catalysis by exchangeable cations, surface catalysis, or the high local concentration of the adsorbed organic molecules.

Clays may have been important on the primitive Earth since they are able to adsorb organic molecules, catalyze their reactions and desorb the reaction products much in the same way an enzyme catalyzes a chemical transformation. Because clays have the capacity to bind a variety of organic compounds, they may have catalyzed a number of different reactions on the early Earth.

Oligomers of amino acids have been prepared in laboratory studies using clays by cycling a clay-amino acid mixture through a wet/dry-heat cycle several times. A degree of polymerization of 2 to 5 has been reported. Catalysis by the dipeptide histidyl-histidine and an enhancement in the proportion of the higher molecular weight oligomers by RNA homopolymers has been observed. Much larger polymers (degree of polymerization up to 60) have been prepared by the reaction of amino acid adenylates on clays. Apparently this polymerization proceeds by the alignment of the monomer on the clay surface; no heat or other activation is required. However, the starting material is a highly reactive compound, and it is not clear that it would have formed in appreciable amounts on the primitive Earth.

RNA oligomers have been prepared by the dry phase polymerization of 2',3' cyclic nucleotides in the presence of clays. Low yields of oligomers with a degree of polymerization as high as 12 have been obtained. These oligomers contain 62 percent of the natural 3',5'-nucleotide linkage and 38 percent of the 2',5'-linked material.

A limited number of studies have been performed using other minerals. The calcium phosphate derivative hydroxylapatite binds higher molecular weight RNA oligomers more strongly than smaller ones. In this way it facilitates the template-directed synthesis of RNA oligomers by adsorbing the larger oligomers while the smaller oligomers remain in the solution phase until they grow to sufficient size to be strongly adsorbed. Studies have also been performed on the iron mineral akaganeite, iron hydroxide soils, and volcanic basalts.

There should be a significant increase in the understanding of the mineral-catalyzed formation of oligonucleotides and oligopeptides by 1995. New findings should include the discovery of polymerization reactions that are more consistent with conditions believed to have been prevalent on the primitive Earth, and the stereospecific formation of polymers, in which the monomer units all have the same chirality. The properties of minerals will have been more thoroughly investigated by 1995, so we will probably have a better understanding of how they may have catalyzed the transformations of prebiological molecules.

The accomplishments of the 1985 to 1995 decade will provide the foundation for investigation of more complex scenarios for the origins of life. Of prime concern is the possible role of mineral catalysis in a system in which the synthesis of polypeptides is coupled with oligonucleotide synthesis. The surfaces of minerals may have served to bind and catalyze the reactions of the primitive translation process. The possible role of minerals in adsorbing and thereby segregating the molecular species essential for life's origins from the complex mix of nonessential organic compounds can then be studied. In other words, it will be possible to investigate experimentally whether minerals served as the skeletal framework on which the origins of life took place.

Lipids and Compartments

A cell is a compartment, and even a rudimentary cell has

an enclosing membrane. We do not know the chemical nature of the barrier membrane of the earliest cells, but some form of lipid is a reasonable candidate, and has the conceptual attraction of continuity; lipid is a major component of the membrane in today's cells. A selectively permeable membrane would confer several benefits on the protocell that used it. It would allow the concentration of essential compounds, energy trapped by the cell could be retained until needed; the internal environment of the cell could be buffered against sudden changes "outside," and, if the cell possessed a unique polypeptide or polynucleotide that, for example, catalyzed more rapid translation or replication than was achieved by the surrounding cells, it would help the cell retain this selective advantage over competitors. In addition, a membrane would protect a set of interacting molecules from dilution by rain or the incoming tide.

Early life was probably based on compartments. This has been realized for some time. Only recently has experimental work indicated that some form of lipid was a likely component of the membranes of the earliest living cells. Thermal cycling of a mixture of fatty acids, glycerol, and inorganic orthophosphate has been found to yield simple phospholipids. Under the conditions of the experiment, the lipid-bilayer compartments known as vesicles formed spontaneously. Fatty acids are not easy to make under simulated prebiotic conditions, but short-chain fatty acids are found in meteorites. Indeed, vesicle-like structures have been seen in aqueous dispersions of nonpolar extracts of the Murchison meteorite. Thus, lipid-like molecules are attractive materials from which to construct primitive membranes: they readily form compartments that are self-healing, their membranes can be made selectively permeable, and suitable molecules could reasonably be expected to have been present on the prebiotic Earth.

By 1995, new pathways for the prebiotic synthesis of fatty acids will probably have been defined. More work will probably be done over the next 10 years on the interactions between lipids and several other classes of compounds. For example, more will be done on encapsulating oligonucleotides and oligopeptides—driven partly by the pharmaceutical industry's interest in using lipid vesicles to deliver compounds to the inside of cells. The effect of lipids on the polymerization of nucleotides and of aminoacids will have been further studied. This work will require detailed analysis with modern analytical instruments. There is present interest in the controlled "replication" of a vesicle. Under what conditions will a vesicle pinch off and divide in two? Could the contents of a vesicle control its size? What happens if additional phospholipid is supplied continuously to the inside of a vesicle? Speculation on how to achieve selective permeability of lipid vesicles will be replaced increasingly by new experimental work. Other wall materials will also be investigated, although they appear inferior to lipids at present.

The use of lipid vesicles in exobiological research is relatively new. Consequently, the status of this area of interest in 1995 is more difficult to predict than for some more established areas. However, the task group expects that the following will be unsolved problems, of great interest to researchers in exobiology:

1. The design and encapsulation of prebiotic light-transducing pigments.

2. The use of these molecules to generate pH or potential gradients, which can then be used to drive chemical reactions.

3. Sophisticated schemes for the polymerization of biomonomers inside a vesicle.

4. Controlled replication of a vesicle, perhaps coupled to polymerization reactions that are taking place inside.

Early Biological Evolution

A major goal of NASA's life sciences program is a greater understanding of early biological evolution. Essential components of this understanding are the historical course of early metabolic and structural diversification, and the ways in which this course was constrained or influenced by the chemical and tectonic evolution of our planet. Conversely, we also want to know how biological evolution has influenced the development of the Earth's atmosphere, hydrosphere, and rocky surface.

Recent research in Precambrian paleontology has shown a hitherto unsuspected degree of interpretable pattern in the early fossil record. Microbiological research has revealed a complementary pattern of phylogenetic relationships among living bacteria and unicellular protists. Our present state of knowledge gives us confidence that continuing research along established and developing lines will result in an extensive geological, paleontological, and microbiological data base by 1995. New techniques developed during the next decade will ensure continued progress during the period 1995 to 2015. These include microchemical techniques for the in situ analysis of individual fossils, microbiological techniques for the culturing and analysis of bacteria that are at present difficult to study; new empirical means of establishing phylogenetic relationships among species, new geochronological tools and the refinement of existing methods for accurate dating of ancient sedimentary rocks, and new discoveries of ancient as well as living microbes.

The task group anticipates that the following research agenda will be important during the early decades of the next century:

1. *The Development of Terrestrial Environments.* Biological evolution on Earth took place in the context of a changing earth surface environment. Indeed, at least some of the major steps in the evolution of the physical environment were biologically induced. Careful sedimentological and geochemical studies of ancient sedimentary regimes will be needed to establish with greater precision the environmental context of early evolution. Such studies should include work on the oldest unmetamorphosed sedimentary terrains, but should not be restricted to them. Analysis of younger Precambrian terrains is necessary to trace the path of evolution and the continuing development of environments, as well as to establish the validity of techniques and approaches before they are applied to geologically difficult Early Archean sequences. Of primary importance will be the elucidation of the early history of atmospheric oxygen and its relationship to major events in crustal and biological evolution.

2. *An Integrated Tripartite Approach to Early Biological Evolution.* Current data obtained through studies of the geological record, microbial phylogeny (based on the sequencing of informational macromolecules), and comparative microbial physiology must be integrated to provide a picture of early evolution. This cannot be obtained from any one of these above sources.

3. *The Development of Models to Relate Data Produced by Agenda Items 1 and 2.* A major goal of NASA is to produce a working model of the biosphere—the interacting geochemical cycles that relate the biota and its environment. Simplified versions of this model have great potential to illuminate early biological and physical evolution on Earth. As a single example, it should be

possible to use this simplified model to explore the effects of continental growth on productivity and, hence, on atmospheric oxygen levels. Biological evolution on Earth cannot be understood until the constraints and influences of our planet's physical evolution are elucidated.

The task group stresses the importance to this research of NASA's continued strong intellectual leadership. No other agency can provide the interdisciplinary framework necessary for biological research on early evolution.

4. *The Search for Evidence of Past Life on Extraterrestrial Bodies.* In addition to the search for extant life in the cosmos, investigations specifically designed to detect evidence of past life on places other than Earth are needed. For example, although it is unlikely that the Mars of today has any life, it is possible that life could have evolved on Mars during its early geological history. On Earth, it is only during the last 650 million years that large, complex life forms appeared, following approximately 3 billion years of microbial dominance. If life evolved on other planets, it seems likely (using Earth, our only known example, as a model) that early stages in the evolution of such life included a long period limited to microbial ecosystems. There is evidence that 4 billion years ago Mars had surface liquid water and a thicker carbon-dioxide-rich atmosphere. That is, 4 billion years ago the surface environments of the Earth and Mars may both have been suitable for the formation of living systems. High priority must be placed on exobiological missions to Mars that call for paleontological and geochemical analyses of samples of ancient sedimentary rocks.

The Evolution of Complex and Higher Organisms

Unlike most other NASA scientific programs, research on the evolution of complex life is not tied to specific space missions or technological developments slated for the next decade. Thus, it is difficult to determine with any degree of accuracy where the field will stand in 1995. For several decades, NASA has played a major role in research on the origin and early evolution of life on Earth; more recently, the agency has assumed a position of leadership in integrated geological, oceanographic, atmospheric, and biological studies of our planet in its present state—the projected "Mission to Planet Earth" (discussed in a separate volume). A sophisticated understanding of our planet's developmental history will require the maturation of both these programs, plus the development of new research programs on the evolution of morphologically complex organisms. In essence, it is this last discipline that links research on the early Earth and the present workings of the planet into an integrated picture of biological evolution through geological time.

There is at present a considerable body of research being funded by the National Science Foundation (NSF) on the evolution of plants and animals. NASA should not duplicate or annex this program. Rather, NASA should seek to develop areas of research that specifically relate to its initiatives in global habitability, the early evolution of terrestrial life, and the possible effects of extraterrestrial phenomena on terrestrial evolution. The task group recognizes three areas of research that fit this criterion. All are poorly funded by other agencies at present, and all represent fields in which NASA has unique capabilities to forge exciting programs of research. A short description of each area follows:

1. *Extraterrestrial Bodies and Biological Evolution.* Traditionally, evolutionary patterns documented in the fossil record have been interpreted in terms of earth-bound processes; however, recent paleontological and geochemical research indicates that there is a high probability that some, perhaps all, mass extinctions were linked to impacting comets or asteroids. Patterns of extinction among marine invertebrates have been interpreted as indicating a 26-million-year periodicity in mass extinction events. Thus, we have the fascinating possibility that animal evolution over the past 650 million years has been controlled in no small measure by periodic mass extinctions whose period is defined by events or processes in the solar or galactic environment. This hypothesis is divisible into several component ideas, each of which is the subject of lively current debate. The outcome of the debates cannot be predicted at the moment, but it is clear that the coming decade will see the development of an extensive body of new data on taxonomic and ecological patterns of extinction during and between mass extinction episodes, as well as on the time scales on which individual extinction events occur. These data will be complemented by astronomical research on extraterrestrial phenomena that could produce periodic patterns. By 1995, it will be time to develop a new generation of models to explain extinctions,

their role in the evolution of complex life, and the role of extraterrestrial events in determining these biological patterns. NASA can and should assume a leadership role in this interdisciplinary effort.

2. *The Origins of Multicellular (tissue-grade*) Organisms.* NASA has led research efforts on the Precambrian origin and evolution of microbial life on Earth; NSF has sponsored much research on the evolution of multicellular life during the Phanerozoic Eon. At present, no coordinated research program exists that is aimed at the interface between the two—namely, the origins and diversification of tissue-grade algae and animals. Such a research program will require the integrated efforts of specialists in a wide spectrum of disciplines, including cell biology, developmental biology, organismic evolutionary biology, molecular phylogeny, paleontology, sedimentary geology, and geochemistry. Questions of importance include the evolution of cellular mechanisms that permitted and facilitated the emergence of organisms with differentiated tissues; the course of early plant and animal diversification as determined from comparative biological and paleontological studies, the impact of new grades of biological organization on ecosystem function; the composition of the atmosphere and hydrosphere; the possible effects of extraterrestrial perturbations; and the possible role of physical Earth evolution on the timing of animal and plant evolution. This last question is of interest with regard to inquiries about the possible evolution of intelligent life elsewhere in the universe (SETI). Is the tempo of biological evolution on Earth controlled biologically, or is it constrained by rates of planetary development? The answer to this question is of immense importance to exobiologists.

3. *The Evolution of the Biosphere.* Major events in biological evolution affect the physical composition and operation of the biosphere, and conversely, important physical changes in the crust, oceans, or atmosphere influence the subsequent course of biological evolution. The geological record provides what is really our only testing ground for models of biosphere function developed as part of NASA's Mission to Planet Earth. Equally important, once global biosphere models become sufficiently sophisticated, it

---

*The term "tissue grade" is used to denote a level of structural complexity characterized by the differentiation of tissues composed of morphologically and physiologically distinct cell types. This is distinct from the multicellularity seen in some algae and prokaryotes in which there is either no cellular differentiation or only the differentiation of individual cells.

should be possible to "run them backward" through geologic time to formulate an integrated view of earth history.

How has life persisted on this planet for nearly 4 billion years? How have biological innovations such as the origin of unicellular and multicellular photoautotrophs, skeleton-forming animals, or wood-producing vascular plants influenced the composition of the atmosphere or rates of continental weathering? What, if any, long-term effects have mass extinctions had on biogeochemical cycles? How have physical events such as the Archean growth of large stable continents, plate tectonic movements, ice ages, or sea-level changes influenced the biosphere? The answers to these and related questions are fundamentally important to understanding the Earth's unique status in the solar system. They are also of tremendous practical importance in that the answers to these questions will determine how and where future exploration for fossil fuels and economically important minerals will take place.

The task group's recommendations are consistent with those presented in the NASA workshop report *The Evolution of Complex and Higher Organisms* (1985), which deals with the establishment and immediate needs of a research program on the evolution of morphologically complex and intelligent life. That document can be consulted for voluminous background information.

NASA is uniquely qualified to lead and coordinate research in this area. Indeed, NASA is uniquely responsible for such coordination in that an integrated model of earth history will be the crowning gem for several of its major initiatives in biology and earth science.

Search for Extraterrestrial Intelligence (SETI)

Are we alone in the universe? The interest in this fundamental question permeates many other areas of space research. Indeed, as *Astronomy and Astrophysics for the 1980's* (National Academy Press, 1982) states, "It is hard to imagine a more exciting astronomical discovery or one that would have greater impact on human perceptions than the detection of extraterrestrial intelligence."

At present, there is a consensus that the best way to try to detect extraterrestrial intelligent life is through a coordinated search for radio signals from technologically advanced civilizations. A complementary approach is based on the technology that has led to the discovery of possible planetary systems beyond the solar system. This includes Van Bisbroeck V-8, a red hot planet or brown dwarf, and the infrared-emitting matter orbiting around beta-Pictoris. This complementary approach uses highly sophisticated infrared detectors and will continue to be pursued for its own sake by planetary astronomers. However, a further refinement and extension of its methodology would be of synergistic value to SETI. Detection of any planets, even if the size of Jupiter, increases the statistical odds that earth-like planets, life, and advanced civilizations may be relatively abundant in the galaxy. If techniques reach the point at which light from an earth-like planet can be investigated spectroscopically, it would be easy to test for the presence of a surplus of atmospheric oxygen. This would constitute evidence for the existence of life, even if that life were very primitive.

But all of these intermediate steps can be bypassed if radio contact with an advanced civilization is achieved. As we write this (January 1986), a proposal is being submitted for a 10-year radio frequency SETI search with state-of-the-art technology. This program will consist of two parts: an all-sky survey in the frequency range 1 to 10 GHz, and a targeted search in the 1- to 3-GHz range. The difference between the two is simply that the targeted search will be a high-sensitivity narrow-band study of selected Sun-like stars. In other words, this search makes the additional assumption that the most likely locations for advanced civilizations are earth-like planets that orbit Sun-like stars. It is the only assumption for which we have any scientific support at present, since we are the only advanced civilization we know. The all-sky survey is free of this assumption. Should a detection be achieved in this mode, the full sensitivity of the system could be brought to bear for confirmation and study.

The enabling technology that is making these two approaches practical for the first time is a multichannel spectrum analyzer and a computer software system that can analyze the flood of data this device will produce. It will now be possible to detect and analyze signals with 10^7 channels simultaneously. This significantly speeds up the work while increasing the sensitivity. The multichannel spectrum analyzer is followed by a pattern detector that searches for regular pulse trains or drifting signals in an array of spectra. This approach will provide a search billions of times more comprehensive, in terms of sensitivity and frequency coverage, than the sum of all previous searches.

Further improvements can be envisaged. Depending on the outcome of this 10-year program, we can anticipate some interest in using very large, filled arrays to pursue the search with still greater sensitivity. We should also be open to other ingenious suggestions for ways to detect advanced civilizations. These might involve searches in other frequency domains of the electromagnetic spectrum.

There seems to be a strong feeling among the space research community that support of such programs by U.S. funding agencies is a legitimate scientific activity, and that choice of programs within each agency should be made through the normal process of peer review. As a number of other nations have initiated steps toward SETI programs, the opportunities for international collaboration are substantial.

The Search for Extraterrrestrial Intelligence (SETI) is a program of exploration currently in an early stage of technological development. Once the engineering design and construction phase of the signal processing equipment is completed, the search, or operational phase, can begin. Funding of an appropriate magnitude will be required to carry out the microwave observing program. This program, while managed by NASA's Life Sciences Program Office, should not compete for funding from other life sciences research resources, but should be funded as a new NASA initiative.

SPACE MISSIONS

Introduction

This section of the report describes the interest within the life sciences community in some highly sophisticated missions to selected objects in the solar system. These missions will advance our knowledge about environments that can test ideas about the origin and evolution of life on Earth, and help to define the qualities that make our own planet so uniquely habitable. Hanging in the balance is the question of the prevalence of life in the universe.

To place these advanced missions in perspective, we must first try to estimate as best we can with scheduling uncertainties, what we will know by the year 1995. An approximate idea of the perspective we will have by that time can be gained simply by reviewing the planetary missions that are already under way or scheduled for the future. Optimistically, the task group has included several missions in this section that are still in the planning stages. These are listed in Table 2.1 and summarized in the next section. The summary is given by target object, with a stress on research goals relevant to the life sciences.

It is important to stress that this suite of missions is largely uninfluenced by the interests of the life sciences community. With a few exceptions, there is little emphasis on the detection and identification of large organic molecules. While the further characterization of planetary environments is obviously necessary and desirable from a biological perspective, the key measurements that will tell the level of complexity reached by chemical evolution or the processes involved in that evolution are usually slighted. All of this adds up to a requirement for dedicated or sample return missions, which are described in a later section.

Pre-1995 Planetary Missions

Venus

The U.S.S.R. has deployed (June 1985) two spacecraft from its Vega project. These involve atmospheric probes with surface sampling capabilities, and the release of balloons to study atmospheric circulation.

The United States plans a radar mapping orbiter to Venus in the late 1980s or early 1990s. This spacecraft will provide a topographic map of the entire planet, with surface resolution approaching 150 m in selected areas. For the life sciences, Venus functions as a control experiment. What would Earth be like if it were closer to the Sun? But is solar proximity the only reason for the extraordinary difference between Earth and Venus? These are two sets of questions these missions are trying to answer.

Mars

The U.S.S.R. has approved a mission called Phobos that is currently scheduled to reach Mars in 1989. While making some reconnaissance observations of the martian surface, this mission, as its name implies, will have the study of the martian satellite Phobos as its main goal. The mission consists of an orbiter that

TABLE 2.1 Missions--Planned or Under Way

Name	Agency (Country)	Launch	Arrival	Target	Objective
Voyager 1	NASA	1977		Jupiter Saturn	Flyby reconnaissance-- images, spectra, magnetic and other fields, particles, etc.
2	NASA	1977	1986	Jupiter Saturn Uranus	
			1989	Neptune	
Vega 1,2	U.S.S.R.	1984	1985	Venus	Atmospheric, cloud, and surface composition; meteorology (includes balloons).
			1986	Halley's Comet	Composition of inner and outer coma, appearance of nucleus, magnetic, and other fields and particles.
Giotta	ESA	1985	1986	Halley's Comet	As for Vega, but Giotto goes much closer to nucleus.
Planet A	Japan	1985	1986	Halley's Comet	Plasma environment and hydrogen cloud around comet.
Galileo	NASA	?	?	Jupiter	Atmospheric probe with in situ composition and structure measurements. Orbiter with complement of remote sensing instruments.
Venus Radar Mapper	NASA	?	?	Venus	Radar map of surface.
Phobos	U.S.S.R.	1988	1989	Mars	Intensive study of Phobos; includes laser vaporization of surface samples.

TABLE 2.1 (continued)

Name	Agency (Country)	Launch	Arrival	Target	Objective
Mars Orbiter Mission	NASA	?	?	Mars	Detailed spectrometric mapping of surface; studies of dust and volatile transport.
Vesta	U.S.S.R.	1992 or 1994	1994 or 1996	Mars	Atmosphere and surface composition, meteorology; additional objectives to be specified.
				Vesta	Characterization by remote sensing of surface composition; evidence of differentiation.
Comet Rendevous/ Asteroid Flyby	NASA	?	?	Comet Temple II	As for Halley's Comet, but with much more time and detail.
Cassini	ESA/ NASA	?	?	Saturn	Remote sensing of planet rings and satellites, direct study of magnetosphere; probe into Titan's atmosphere.

will match orbits with the satellite, slowly moving past it while irradiating it with a powerful laser and an ion beam. The material ejected from the satellite will be analyzed with mass spectrometers aboard the spacecraft. Numerous other remote investigations of Phobos are planned, and inclusion of a penetrator that lodges in the surface of the satellite is under consideration.

The United States is planning a Mars Orbiter Mission (MOM). This spacecraft will go into a polar orbit about the planet, and carry out an extensive series of mapping studies designed to elucidate the surface composition, the general circulation of the atmosphere, and the behavior and transport of dust and volatiles. MOM is an essential precursor for a Mars sample return.

The relevance of Mars to the life sciences is well known. While it is extremely unlikely that there is any life on Mars today, it remains the only planet on whose surface we have found evidence for the presence of liquid water sometime in the past. This evidence leaves open the possibility that life may once have originated on Mars, only to die out as the global climate deteriorated.

The Phobos mission will provide compositional information on a dark, low-density satellite that is thought to be rich in organic materials, and perhaps representative of asteroidal parent bodies of the carbonaceous chondrite meteorites.

The Soviets have also approved a mission called Vesta that will put a 200-kg payload in orbit around Mars in the 1990s. This spacecraft will deploy at least one penetrator, perhaps into one of the polar caps. There will also be two balloons carrying cameras, gamma-ray spectrometers, and an array of meteorological instruments. An auxiliary spacecraft (currently allocated to the French) will fly on to the asteroids.

Comets and Asteroids

In 1985, ESA launched its Giotto mission to Halley's Comet. It was preceded at Halley by the Soviet Vega spacecraft. Both missions encountered Halley in March 1986. Both of these projects are altering our ideas about comets in profound and fundamental ways. Other spacecraft will study cometary plasmas, but Giotto and Vega have already provided the first pictures of a comet nucleus and will produce the first close-in measurements of composition. The latter will include mass spectrometric studies of the inner coma, searching for the so-called parent molecules of the coma that are expected to include some interesting organic compounds. At this writing (late March 1986), it is already clear that the nucleus of Halley's Comet is covered with a dark, presumably carbon-rich coating, which is probably contributing to the evidence for a high abundance of organic compounds in the inner coma.

The United States is planning a Comet Rendezvous Asteroid Flyby (CRAF) mission. If the mission is approved, the spacecraft will rendezvous with Comet Temple II in the late 1990s, having encountered one or more asteroids en route. Unlike the Halley fast flybys, this spacecraft will stay in the vicinity of the comet nucleus for months, allowing repeated measurements and studies of temporal variations of the release of gases and subsequent chemical reactions. The composition of the nucleus, including studies of the non-volatile components, is one of the prime mission goals.

The Soviet Vesta mission, as its name implies, will go on to the asteroid Vesta in the 1990s, after dropping the aforementioned probes at Mars. The U.S. Galileo mission (see Jupiter below) will also visit an asteroid en route to its planetary target.

Outer Planets

The U.S. Voyager 2 spacecraft encountered Uranus in January 1986 and will visit Neptune in August 1989. This means that every planet except Pluto will have been visited by spacecraft by the year 1995, the beginning of the 20-year period upon which this study focuses. While the lower atmospheres of both Uranus and Neptune are warm enough for some potentially interesting chemistry to take place, the Voyager spacecraft are not equipped to explore these regions. On the other hand, the satellites and rings of Uranus are suspiciously dark, suggesting that they may be coated with carbon-rich material. Furthermore, the dark material on the surfaces of these objects is distinctly different from the dark coatings on Saturn's satellites Phoebe and Iapetus. There is clearly some variety in the types of organic compounds that are currently stored in the outer solar system. Neptune's largest moon Triton may have oceans of liquid nitrogen on its surface as well as a methane-containing atmosphere.

The Galileo mission to Jupiter will arrive sometime in the 1990s. It consists of an orbiter with a probe that will be deployed for measurements in Jupiter's atmosphere. The atmospheric composition will be determined to a sensitivity of better than 1 ppm for many constituents. Without a gas chromatograph, however, these measurements, will just be the first step in an in situ characterization of trace organic molecules.

A similar mission for the Saturn system is currently in the planning stage. Called Cassini, it would be a joint project between ESA and NASA, in which the Europeans would build a probe to be sent into the atmosphere of Titan, while the United

States would construct an orbiter that would include a radar mapping instrument for studying the satellite's surface. Because the atmosphere of Titan is already known to contain a rich variety of organic molecules, the probe is being designed to include a gas chromatograph. Current plans also include the possibility of some post-impact measurements to try to characterize materials on Titan's surface.

Planetary Missions After 1995

Sample Return Missions: The Next Step

Assuming all of the missions described in the previous section are approved, launched, and successfully fulfill their objectives, the new perspective we will have in 1995 can be summarized as follows: Spacecraft will have visited every planet in the solar system except Pluto. The Space Telescope will have given us our first clear views of Pluto and its unusually large satellite Charon. At least one of the two dark, asteroidal moons of Mars will have received a detailed investigation, and at least three bona fide asteroids will have been surveyed. A wealth of information about Halley's Comet will have been assimilated, and data on the behavior of Comet Temple II may shortly be arriving from a spacecraft poised in its vicinity. This (CRAF) mission will be equipped to deliver much more detailed information about the organic material found in cometary nuclei than the Halley missions could tell us.

The existing suite of in situ measurements on Venus may have been extended in number and type. Detailed topographic maps of Venus and geochemical maps of Mars may allow an intercomparison of the physical and chemical processes at work on all three large inner planets at an unprecedented level of detail. The information from the Galileo Jupiter probe may provide a fundamental calibration scale for elemental and isotopic ratios made throughout the solar system. Galileo will also provide the first direct investigation of a highly reducing planetary atmosphere and the chemistry that takes place there. A mission may be under way to bring our knowledge about Saturn's satellite Titan up to this same level of understanding. This mission should provide far more information about organic chemistry in a reducing environment than Galileo could.

To move forward past this point, it is necessary to consider a much more ambitious class of missions: those that involve the return of samples to Earth. It is sobering to realize that this most difficult enterprise was first accomplished in 1969 when Armstrong, Collins, and Aldrin returned to Earth with material from the Moon. Twenty-six years later, the Moon will remain the only celestial object from which we have collected samples. This circumstance may be attributed primarily to the high cost of such missions. Given the sophistication of modern analytical techniques, is it really necessary to pay this price? Do we really have to bring samples back to our earthly laboratories?

There is a clear consensus in the planetary community that the answer to both of these questions is affirmative. Among the unique data that can be obtained from returned samples are those described in the following list (based on the Solar System Exploration Committee (SSEC) report titled *Planetary Exploration Through Year 2000* (SSEC/NASA Advisory Council, 1983), describing augmented missions):

1. *Independent Ages and Historical Data.* These are determined from highly precise measurements of radioactive isotope parent-daughter pairs in selected samples and different fractions of the same sample. Such measurements require large mass spectrometers and "clean-room" techniques. The resulting ages permit the timing of planetary events and the time limits on processes during the early solar system.

2. *Precise Chemical and Isotopic Data.* Remote measurement techniques cannot compete with the abilities of terrestrial laboratories to measure abundances of 60 to 70 elements to levels of parts per billion, including precise isotope ratios.

3. *"Ground Truth" for Global Planetary Missions.* Once we have detailed information on a few samples, we can use this to calibrate the global mapping data available from precursor orbiters.

Among the operational advantages accruing to the analysis of returned samples, the following can be mentioned:

1. There are virtually no limits on the instrumentation that is available to study the samples. Given enough material, virtually any type of analysis can be performed.

2. The analytical instruments being used are state-of-the-art. They do not suffer from the 5- to 10-year lag that is common in the case of space missions.

3. There is greater planning flexibility. No assumptions need to be made in advance about the nature of the sample to be tested, about its physical state nor its composition.

4. Analyses are open-ended and not time-limited. They can be repeated (even after long time intervals), done in variable series, and repeated by various investigators in different laboratories using a variety of techniques.

Given all these advantages, it is not surprising that the community of scientists interested in the planets is actively promoting sample return missions as the next step in planetary exploration in the years 1995 to 2015. In later sections, two of these missions—sample return from Mars and from a comet nucleus—will be considered in some detail.

Targets and Strategies for Future Missions

The purpose of this section is to define the next large steps forward in solar system exploration beyond the year 1995 that would be most important to the life sciences. We have seen that by 1995 it will be possible to design sample return missions to several objects. Yet other targets, especially Titan, will require additional exploration of a highly sophisticated nature before sample return can be attempted. The task group will consider both sample-return and exploratory missions in this report.

The distinction between the two is well brought out by the criteria established for targets of sample return missions set out by the SSEC. For sample return missions:

1. Targets must be accessible with the technology available in the relatively near future.

2. Collecting devices and associated hardware must be able to survive and function on target surfaces.

3. Targets must be understood well enough as a result of previous missions that a productive sample return mission can be planned in adequate detail.

The task group would add a fourth criterion:

4. Targets must show evidence of chemical evolution relevant to the problem of life's origin—either as processes taking place today or in the form of deposits of products from earlier epochs.

Using just the first three criteria, the SSEC selected Mars and a comet nucleus for their candidate missions, to be initiated prior to the year 2000. These two targets nicely satisfy the fourth criterion as well.

As the SSEC report states in the case of Mars, "Important bioscience goals could also be achieved through a sample return. The presence or absence of indigenous life—especially fossil life—and the reasons for the surprising absence of organic matter in the martian soil could be determined directly by the laboratory examination of an unsterilized sample returned to Earth." The report points out that comets ". . . are an entirely new and virtually unstudied class of objects which, because of their primitive nature, are fundamentally important to understanding the origin and earliest history of the solar system." With a returned sample, "a complete characterization of cometary organic material would be possible, with exciting implications for understanding the origin of life. . . ." Mars and the comets are at the top of the task group's list, too, and it strongly endorses the SSEC recommendations that these be the first of the sample return missions.

Given that the task group's study of future missions extends 15 years beyond the limit accepted by the SSEC, the task group can legitimately add some missions to the SSEC list. These might include precursors to sample return missions such as soft landers and penetrator networks. Titan is certainly a prime candidate for a soft lander mission, similar to the Viking project that studied Mars in 1976. We will need the results from the Cassini mission before such a mission can be planned in detail. At that point we will know whether we must use a lander that floats or whether a more conventional spacecraft will suffice. It will also be possible to address such issues as the location of the landing site and the type of surface sampling devices.

There has also been great interest in a possible mission to Europa, the smallest of Jupiter's Galilean satellites, which may be covered with an ocean of liquid water whose surface is shielded by an icy crust. Here, the next step in exploration after the Galileo mission will be the deployment of some type of seismic network to determine whether the subsurface ocean really exists. The penetrators that carry the seismometers could also be equipped with devices to make some compositional analyses of their immediate surroundings. Thus, it would be possible to determine, for example, if the dark material associated with some of the "cracks" on the surface indeed consists of organic compounds as some have suggested. Plans for a sample return mission or a subsurface probe could then be developed after the results from this precursor mission were analyzed.

It should be evident that the specific missions being considered here involve techniques for gathering data that can easily be applied to other targets as well. Thus a comet nucleus sample return mission, once it has been successfully carried out for a given comet, will allow the study of many other comets with the same payload and mission strategy. Modifications in the payload would permit the use of the same or similar spacecraft to carry out sample return missions to selected asteroids.

The penetrator network described for Europa could be deployed on asteroids and other satellites. Phoebe and the dark side of Iapetus would both be fruitful targets for such missions. Both of these satellites in the Saturn system have surfaces that seem to be covered with carbon-rich material. The discovery by Voyager 2 that the dark material on the satellites of Uranus has distinctly different optical properties from the material found in the Saturn system invites direct exploration of the Uranus system as well. Given the low temperatures in this part of the solar nebula, this organic material could be even less modified than that found in the carbonaceous chondrites.

The next two sections of this report will concentrate on the Mars and comet nucleus sample return missions. The SSEC studies will form the background for this discussion, which will emphasize those aspects of these two missions that are especially interesting to the life sciences. The final section will cover the soft lander and penetrator missions to outer planet satellites.

Sample Return from Mars

In addition to the high priority awarded it by the SSEC, a sample return mission from Mars was also implicitly endorsed by the Space Science Board of the National Research Council 1978: "The study of Mars is an essential basis for our understanding of the evolution of the Earth and the inner solar system. . . . *We recommend that intensive study of Mars be achieved within the period 1977-1987*" (*Strategy for Exploration of the Inner Planets: 1977-1987*, National Academy Press, 1978).

It is now clear that this recommendation will not be satisfied in the stated time interval. As Table 2.1 indicates, the first U.S. mission to follow the success of the Viking project will be the Mars Orbiter Mission (MOM). Nevertheless, sentiment is building to support a Mars-intensive program that would culminate in the return of one or more samples. This might be done as a cooperative project with the Europe or the U.S.S.R., continuing a welcome international participation in planetary exploration. The sample return missions can also be viewed as precursors for the still more advanced idea of manned Mars missions, ultimately leading to the establishment of a base on the martian surface.

The return of samples from another planet is obviously a very expensive proposition. In the case of Mars, a large number of different scientific inquiries can be aided by the ability to carefully study preserved samples that are successfully returned to terrestrial laboratories. This means that great care must be exercised in the choice of landing site and sampling strategy in order that the greatest number of scientific objectives can be met. It will be necessary to return a variety of carefully chosen samples taken from sites that have been selected with specific objectives in mind. This means that the sample identification and selection devices must be mobile; some type of intelligent rover will be required.

Even with careful planning, there will be conflicts. From a geological point of view, the key units for sampling in order of decreasing priority are young volcanic units, intermediate-age volcanic units, ancient cratered units, layered units, polar units. The last two have the highest priority from a biological perspective. Areas where major sedimentation has occurred may provide the best opportunity for collecting samples that contain evidence for past life on Mars. Of special interest are those regions where standing or slow-moving water is thought to have covered the surface. It may well be possible to combine a sampling expedition to an area of sedimentation with one of the higher ranking geological objectives. But investigation of the polar regions will require a separate mission.

The poles are important for several reasons. The north pole is already known to be covered by a permanent water ice cap. The long-term availability of water even in the form of ice at the planet's surface could have been essential for any surviving indigenous organisms. Both poles serve as cold traps on the planet. They are known to be regions of aeolian deposition; layered terrains have been observed in both locales. These are therefore regions where we might expect primitive organic material to be preserved, protected from the hostile surface environment by layers of ice. To reach this material, it will be necessary to drill through the permanent cap. The missions being designed for Mars sample return include the capability of landing anywhere on the planet. The rover will be equipped with a drill, a core tube, and a sampling arm, all of which could be put to good use on a polar cap mission.

Finally, it is important to stress that many of the geological studies, especially those aimed at an understanding of the history of water and surface-atmosphere interactions on the planet, will be of great interest to the biological community.

The troublesome problems of contamination of Mars and back contamination of Earth require additional study. Both biologists and geologists agree that the samples returned to Earth should not be sterilized, since the required heat or chemical treatment could have a seriously deleterious effect on the samples. The Space Station might play a role here by providing a specialized environment in which preliminary analyses of unsterilized samples could be carried out before shipping them to Earth. One way of using this facility would be to sterilize a small, carefully selected portion of a sample that could then be shipped to Earth for detailed study. The major portion of the sample would remain on the Space Station, presumably undergoing tests for dangerous biological or chemical activity.

This approach will require careful thought and international agreement on the kinds of tests to be run before the samples are considered "safe." Biological safety could be assured in a two-step process in which a preliminary set of tests are run in orbit, while a more sophisticated set would be carried out in laboratories on Earth. This two-tiered procedure would reduce the cost associated with a fully-equipped laboratory on the Space Station.

It is not possible to look beyond sample return missions with much certainty. The discovery of viable organisms in returned martian samples would have a profound impact on strategies for future exploration of the planet. In the more likely case that the samples are sterile, it still seems inevitable that momentum for a manned mission to Mars will continue to grow. The maintenance of the people sent on such a long mission, including their survival on Mars, is another aspect of the problems confronted by specialists in space medicine, as described elsewhere in this report.

Comet Sample Return

Comets potentially offer the least-altered samples of solid material still available in the solar system. This is demonstrably true of substances whose volatility is equal to or less than that of carbon dioxide, and probably holds for still more volatile compounds as well. A comet nucleus has been likened to a dirty snowball, where the snows are now known to consist predominantly of ices of water and carbon dioxide, and the dirt includes carbon-rich compounds in addition to silicates. Here then is an opportunity to examine some of the solid materials from which the planets accreted, unchanged since the solar nebula first formed. Comets represent the solar system's starting conditions, providing examples of the kind of chemistry that went on before there were planets. This includes examples of materials that have been continually delivered to the surfaces of planets—sometimes catastrophically—but sometimes yielding starting materials for new chemical reactions. It is by no means established that organic material from comets contributed to the famous primordial soup from which life on our planet arose. Yet this is one of several current hypotheses regarding the first steps for the origin of life, and it *is* certain that comets must have contributed some fraction of the volatile elements that dominate the chemistry of life as we know it: the hydrogen, oxygen, carbon, and nitrogen from which we are mainly composed.

These considerations make the investigation of comets of great interest to the life sciences. Accordingly, the task group enthusiastically endorses the recommendation of the SSEC that "the return of a sample from the nucleus of a comet is one of the highest priorities for an augmentation mission and should be undertaken as soon as possible."

It is only by bringing samples back to our laboratories on Earth that we can perform the full array of experiments with the finesse required to tell us what we want to know. What compounds are present and in what proportions? Are they really pristine, or is there evidence of some processing during the comet's lifetime? How does this suite of organic compounds compare with those found in carbonaceous chondrites? in the interstellar medium? in laboratory experiments? Are there clues from isotope ratios or rare gas abundances that can let us decide what fraction of terrestrial volatiles were actually brought to Earth by these icy messengers? Do comets represent frozen primordial soup? That is, would a melted comet constitute a good starting point for further chemical evolution leading ultimately to life? These are just some of the questions we would like to answer.

The Choice of a Suitable Comet. As the news from the Halley observations by spacecraft and Earth-based techniques trickles in, it is clear that early ideas suggesting that comets contain substantial amounts of organic material are indeed correct. The special attraction of comets arriving in the inner solar system for the first time is that they may have been maintained at very low temperatures (less than 20K) since the solar system began. Even a comet like Halley's, which has been trapped in a short-period orbit that forces repeated visits to the vicinity of the Sun, displays the full range of phenomena and the variety of molecular fragments that are found in "fresh" comets.

In choosing a comet for a sample return, however, we will have to pick from less active objects that are in orbits of still shorter periods than Halley's majestic 76 years. That means they are, on average, much closer to the Sun and will therefore have lost some of their most volatile constituents. Nevertheless, we know from ground-based spectroscopic observations that there are comets even in this category that are actively emitting gas in which molecules such as CN, C_2, and C_3 can be identified.

Furthermore, the icy nuclei of all comets whose nuclei have been directly observed are found to be very dark. They do not exhibit the high reflectivity associated with the ices we know they contain. For example, the nucleus of Halley's Comet was found to be nearly twice as large as the size calculated from its assumed reflectivity, because it was actually much darker than had been thought. Evidently, these comet nuclei are coated with dark material, which apparently becomes concentrated as the ices sublime away, much as lag gravels are left behind during aeolian erosion. With reflectivities below 5 percent, this material must contain a large amount of carbon. Hence the surface layers of a short-period comet represent a concentrated sample of the organic material that all comet nuclei presumably contain. Thus, it may be concluded that we should be able to answer many of our most pressing questions by sampling one of the small, short-period comets that are relatively common and easy to reach. To make this enterprise worthwhile for scientific purposes, however, we must require that

(1) the comet is still actively giving off gas, and (b) the gas shows the presence of C_3, C_2, CH, and CN radicals.

Mission Description. "It is now technologically within our reach to collect pristine samples directly from active comets and to return them to Earth for intensive laboratory study."

This bold statement from the SSEC report sets the tone for the mission. While bringing back samples from the surface of a comet nucleus is obviously of great interest and must be done, we also want to obtain a deep core sample. This offers the best opportunity to obtain material that has undergone only minimal heating, outgassing, and exposure to radiation that would lead to alteration of the original state of the material.

The extent to which this is possible will depend on the particular comet that is sampled. It is likely that we shall have to sample several nuclei of different comets that are at various stages of disintegration (devolatilization) before we will know just how primitive (or representative) is a particular sample. But even if the most volatile substances such as carbon monoxide, molecular nitrogen, methane, and argon are deficient or absent, we shall be very pleased to be able to obtain samples containing cyanides, aldehydes, and the other more complex substances that are responsible for the dark coatings on these icy nuclei.

The basic mission profile as currently envisaged is as follows:

1. Rendezvous with the comet.

2. Spend some time observing the comet's behavior and selecting appropriate sampling sites.

3. Obtain at least two separate core samples approximately 1 m long and 6 cm in diameter.

4. Deploy a long-lived surface lander to be left on the nucleus.

5. Return to Earth with the samples contained in an environmentally controlled capsule.

This may sound straightforward, but there are a number of serious technical concerns to address before such a mission can be realized. The first of these is the propulsion system. This mission requires a low-thrust capability that is currently (1986) not available. An example is the solar powered ion drive, under study in both the United States and West Germany. An engine of this general type could provide a low, but continuous propulsion that would allow the spacecraft to carry the heavy payload required on the round trip to the nucleus and to carry out the necessary rendezvous and reconnaissance maneuvers at the target comet.

A second major problem is the hazard presented by dust that is expelled by the comet nucleus along with the subliming gas. Both the Vega and the Giotto spacecraft were damaged by dust from Halley's Comet; this was a much greater problem than had been anticipated. The comets selected as targets for sample return will be far less active than Halley, however, and this problem needs to be reevaluated after the CRAF mission has successfully visited one of these better behaved candidates.

Finally, the questions of sample handling and contamination problems will have to be studied carefully. Despite familiarity with extraterrestrial samples through the Apollo program, the low temperature, highly reducing, volatile-rich environment of a comet nucleus will provide new challenges that should be anticipated as part of the mission planning. It would be ironic indeed if the effort and expense of a comet nucleus sample return mission were wasted in the end because of an inability to examine the sample properly on Earth.

In Situ Studies of Titan

On the assumption that the Cassini mission (or some surrogate that accomplishes the same objectives) will have made a successful study of Titan by the year 2005, we can ask what the next stage of exploration of this satellite would be. This assumption implies that we will know the following:

1. The location and extent of lakes, seas, or oceans of liquid hydrocarbons (principally ethane) on the surface of Titan.

2. The composition of the atmosphere to the level of 1 ppm (better in some cases), including isotope ratios for abundant elements and identifications and abundances of volatile organic compounds.

3. A first-order characterization of the chemical composition of aerosols collected during descent.

4. The temperature and pressure profile of the atmosphere along the descent trajectory and the vertical distribution of clouds along this same path.

5. A crude characterization of the surface at the probe's impact site. (This may be no more than a test of whether the surface

is solid or liquid, but it might include some information about the composition of that solid or liquid).

6. A variety of other information—about atmospheric winds, the inorganic composition of the surface, constraints on internal structure, and so on—of less immediate relevance to the task group's specific concerns.

The next step in exploration will require some type of soft lander with a capability to move around on the surface. Depending on the nature of that surface, this could be either a rover, a boat, or some type of amphibious device. An alternative approach would be the use of a powered dirigible with the capability of sampling the surface by means of one or more instrument packages lowered on a tether.

With so much uncertainty remaining about the nature of the environment we wish to explore, it is not possible to be much more specific than this. It seems highly probable from our current perspective that Titan will have *some* liquid hydrocarbons on its surface. We will surely want to examine these—determining their composition, perhaps looking for evidence of optical activity in the organic molecules dissolved in them, scouting their shores for concentrations of materials deposited from the atmosphere. If there is a global ocean the need for *surface* mobility is less obvious, but we would still like to be able to investigate the atmosphere over a wide range of latitudes. The dark polar cap and the hemispherical asymmetry in the reflectivity of the smog layer observed by Voyagers 1 and 2 imply that different chemical reactions are occurring at different locations in the atmosphere. This is borne out by the spectroscopic evidence for latitudinal gradients in the abundances of hydrocarbons and nitriles.

At the present time, we have no specific information about the chemical composition of Titan's aerosols. We only know that they must be a combination of condensates of the volatiles detected spectroscopically and polymers produced by the irradiation of these polymers. The Cassini probe is being designed to have some capability for collecting and analyzing these aerosols, but only along the single probe entry trajectory. To investigate latitudinal differences in chemistry, a very ambitious mission will be called for, but it is simply too early to define it.

The same impediment interferes with efforts to specify a payload. The main thrust of this mission will be to understand the chemical reactions taking place on Titan. What processes are involved and what are they producing? Are there any preferred pathways toward complexity? Are any catalysts available and what role do they play? Do the molecules produced show any specific optical activity? Have there been any changes in these processes over geologic time? Are further reactions occurring at the surface in addition to those in the atmosphere? What relevance, if any, does this history have to the organic chemistry on the primitive Earth?

Attempts to answer these questions (and others like them) will require an array of sophisticated instruments operating in a benign environment that is maintained in spite of ambient temperatures of 94K or less. This will be a formidable challenge, but one that surely can be met. The opportunity to explore this natural laboratory (and repository) already seems well worth the effort. The task group anticipates that the results from the Cassini mission will simply increase the appeal of this next step in the exploration of Titan.

CONCLUSIONS AND RECOMMENDATIONS

- The set of missions described in Table 2.1 implies that significant advances in our understanding of the various members of the solar system will be achieved by the year 1995.
- The Task Group on Life Sciences particularly endorses the new missions to Jupiter, to comets, to the martian satellite Phobos, to Mars itself, and to Titan. But the task group is concerned that most of these missions are being planned with little concern for the detection and analysis of organic compounds with high molecular weights. The task group strongly urges the spacefaring countries of the world to give greater consideration to this issue.
- Mars, comets, and the satellite Titan are currently the prime targets for intensive exploration from the perspective of the life sciences. The task group strongly supports efforts to learn more about the progress and history of chemical evolution on these bodies. The dark asteroids and the dark material found on the surfaces of icy satellites in the Uranus and Saturn systems also merit further investigation from this same point of view.
- Some specific recommendations for continued planetary exploration are as follows:

1. *Mars*: The task group recommends intensive exploration of the surface to define regions most suitable for subsequent in situ analysis. These areas include layered terrains such as those exposed on canyon walls, areas of sediment deposit such as the floors of ancient basins that held standing water, and the ground around and under the polar caps. This exploration phase is then to be followed by the deployment of highly instrumented rovers or the return of samples for analysis on Earth. *The choice of samples to be brought back from Mars must include an assessment of their relevance to the problem of life's origin.*

2. *Comets*: The task group recommends further characterization of the compositional and behavioral heterogeneity of comet nuclei and evaluation of the dust hazards in their immediate vicinities. This should be followed by sample return missions that will bring unaltered samples to Earth for detailed chemical analysis. Concurrently, there must be development of facilities to handle these low-temperature, volatile-rich samples.

3. *Titan*: The task group recommends that the initial orbiter-probe mission (Cassini or its surrogate) should be followed by a mobile lander (or a floater with surface-sampling capability) that can carry out sophisticated analyses of the organic materials accumulating on Titan's surface and define the processes that lead to their production.

3
Global Biology/Biospheric Science

BACKGROUND

Over the last two decades we have become aware that the traditional division of the sciences related to the study of the Earth—biology, oceanography, geology, meteorology—represents only an historic artifact inherited from eighteenth- and nineteenth-century scientific thinking. We are learning that the processes that characterize the Earth's environment transcend these boundaries, and that the traditional "spheres"—biosphere, atmosphere, hydrosphere, geosphere—are parts of one complex system. As a consequence, we are witnessing the development of a new science, the science of Earth as a system—a discipline that as yet has no proper name. It is this new, unified science that the task group will address in this chapter with the (inadequate) terms "global biology" and "biogeochemistry."

The perspective from space, which made it possible for the first time to capture in one glance the entire planet—with its continents, oceans, atmosphere, and biosphere, has contributed significantly to this growing conceptual synthesis. Because of this inherently global perspective, the task group feels that space science should develop and maintain a leading role in global biogeochemistry. The current divisions along the lines of traditional sciences reflect themselves in the fact that, during this study, global biogeochemistry was addressed both by the Task Group on Earth Sciences and the Task Group on Life Sciences. To resolve this problem to some extent, the two groups have coordinated their approach. The Task Group on Life Sciences strongly endorses the concept of the "Mission to Planet Earth" discussed in a companion volume by the Task Group on Earth Sciences. In the current report, the Task Group on Life Sciences will limit itself to the discussion of the fundamental concepts that it feels will make up the basis for the space science effort in global biology and biogeochemistry in the period 1995 to 2015. For details of research planning and implementation it will refer in most cases to the Task Group on Earth Sciences report. Much of the conceptual background that forms the basis of the discussion in this report has been covered in two recent NRC reports: *A Strategy for Earth Science from Space in the 1980s and 1990s* (National Academy Press, 1985) and *Remote Sensing of the Biosphere* (National Academy Press, 1986). The document describing the Earth Observing System (EOS) (NASA T.M. 86129, August 1984) also provides much of the scientific background for the topics discussed here, as well as a description of the types of sensors the task group expects to be in orbit by 1995. The task group has considered the existence of an EOS-type system as the starting point for its projections.

Before addressing the directions for space science in some specific areas—biosphere-atmosphere interactions, biogeochemical cycles, global ecology, evolution of the biosphere—some points that are fundamental for space science planning beyond 1995 must be discussed: sensor traceability, information nesting, and data handling.

The detection of long-term trends (time scales of decades to centuries) is essential for our understanding of the response of the earth system both to the large-scale perturbations brought about by human activities and to the effects of geophysical changes. To detect such trends, the calibration of instruments and sensors has to be maintained over long time periods. When sensors are being replaced by new ones, the characteristics of old and new sensors need to be compared carefully. For example, the importance of long-term calibration traceability has become evident in the history of atmospheric carbon dioxide measurements, where the careful maintenance of calibration procedures has allowed the unequivocal detection of historical trends. As a contrasting example, the lack of such traceability still obscures the existence of clear trends in acid deposition.

The concept of information nesting requires that information obtained using space-based sensors be embedded in ground- and aircraft-derived data so that a continuity of the scales of observation can be obtained. Space observations have to be linked to ecologically and geochemically relevant variables through careful calibration with ground-based observations—so-called ground truth. This program must be integrated from the start in space research activities; it should not be assumed that the necessary ground- and aircraft-based activities will be funded and undertaken by agencies other than NASA unless this has been coordinated in advance.

Finally, the volume and character of the data stream that is expected to result from the deployment of a variety of sophisticated space platforms and sensors, and from the associated ground and aircraft activities, will require the development of new data handling capabilities. This development should precede the deployment of the data sources, such as satellites. Too many data exist already that are practically inaccessible due to neglect of data handling needs. The development of complex analytical and numerical models will be necessary; these models will have to be equipped to receive and process the multitude of ecological, meteorological, chemical, and geological data that will be available from the instruments suggested in this and the Task Group on Earth Sciences report. The requirements for data handling or modeling are described in more detail in a subsequent section.

BIOSPHERE-ATMOSPHERE INTERACTIONS

The biosphere and the atmosphere are intimately connected: the composition of the atmosphere is largely a result of the activities of the biosphere; on the other hand, the chemical and climatic characteristics of the atmosphere are essential for the support of life on Earth. Due to its relatively small mass compared to that of other components of the earth system, the atmosphere responds rapidly to changes in the input and output of both material and energy. This makes it valuable as a "fast-response sensor," but also makes it very susceptible to environmental impact from human activities. In the following sections, the task group will discuss the needs for space research related to the interactions between the biosphere and global climate and atmospheric chemistry.

Biosphere-Climate Interactions

In this century, mankind began to conduct two very large, albeit inadvertent, experiments on the interactions between the biosphere and climate: the addition of carbon dioxide, and some other "greenhouse" gases to the atmosphere, which is expected to raise global temperatures, and the large-scale deforestation of the tropical rain forest regions. These "experiments" will reach such an advanced stage in the timeframe considered in this report that we can expect a clear climatic and biospheric response.

Most of the trace gases that contribute to the greenhouse effect (carbon dioxide, methane, nitrous oxide, etc.) have biospheric sources and sinks as well as anthropogenic ones. It is essential that we understand the processes leading to their emission and removal, as well as their fluxes to and from key environments. This will require a combination of ground-based aircraft and space platform studies, which can be developed on the basis of the experience gained in experiments like the NASA GTE/ABLE and EOS projects. In conjunction with the activities in atmospheric chemistry discussed below, this research should lead to a model of atmospheric trace gas chemistry that would have predictive capabilities and that could be linked to climatic models in order to investigate the climate response to the expected combination of greenhouse gases. Currently, carbon dioxide is considered almost exclusively by atmospheric modelers, although the effect of other gases could be as important.

Atmospheric aerosols have climatic effects that depend for their direction and magnitude on the chemical and physical characteristics of the aerosol. We need an improved understanding of the role of the biosphere in the production of aerosols and aerosol precursor gases, such as reduced sulfur gases. It is likely that biogenic aerosols have a significant influence on the albedo of oceanic clouds; we need to study this relationship using aircraft and satellite techniques. The effect of marine cloud albedo on the Earth's radiation balance is significant and requires further study.

The large-scale change in surface characteristics due to tropical deforestation, desertification, and changing agricultural practices must be expected to influence substantially the transfer of both heat and water vapor at the Earth's surface. The mechanisms and rates of energy and water exchange as a function of ecological and soil characteristics must be further investigated using a combination of active and passive remote sensing (microwave, infrared, visible) with ecological ground studies. A very important role could be played here by sensors that are deployed at ground level and which transmit their information to satellites, which in turn relay this information to data analysis centers. Remote sensing must be employed to keep track of the changes in extent and distribution of biogeographic zones. All these data will have to be integrated into climatic models to predict the response of the atmosphere. Conversely, as the effects of climatic change become apparent in meteorological observations, we will have to look for responses of the biosphere to these changes. These responses may not be readily detectable at ground level due to the large local variability of ecological parameters, but may be evident from the much larger data set accessible through remote sensing. A prerequisite for such studies will be a much enhanced ability to determine ecological variables by remote sensing and to process such data to a large extent by automated techniques. The Task Group on Earth Sciences report contains a detailed discussion on the relationship between the global hydrological budget and the biosphere.

Largely as a result of satellite data, our understanding of ocean-climate interactions is steadily improving. The sensitivity of high-latitude climate to perturbations by the "greenhouse" effect and the importance of high-latitude regions in oceanic productivity make space-based studies of the interactions between oceanic productivity and climatic change—especially at high latitudes—an essential topic for future investigation. Most of the transport of surface water to the deep ocean—and consequently the removal of carbon dioxide to the deep waters—also occurs at high latitudes. Due to the difficulty of observation from ships, with regard to both spatial and temporal coverage, space studies will play an essential role here. We will require space sensors to determine pigment concentrations and to deduce oceanic primary productivity and (as far as possible) phytoplankton types. We will have to look for changes in marine biogeography and ecology in response to climatic change. From these studies we hope to gain not only fundamental information on global marine ecology, but also essential input to the management of marine resources.

Biosphere-Atmospheric Chemistry Interactions

The role of the biosphere as a source and sink for most atmospheric constituents has already been mentioned. Biospheric processes are essential for the atmospheric content of carbon compounds (CH_4, CO_2, CO, alkanes, etc.), nitrogen species (NH_3, NO_x, etc.), sulfur compounds (H_2S, $(CH_3)_2S$, H_2SO_4, COS, etc.), organohalogen compounds (e.g., CH_3Cl), and many other molecules. For a number of these compounds, spaceborne sensors exist or are under development, for example, in the UARS (Upper Atmospheric Research Satellite) project. However, the use of these sensors is largely limited to the upper atmosphere. It will be essential to develop techniques for remote sensing of tropospheric constituents. These instruments will include active sensors, and will be operated from satellites and, to a large extent, from research aircraft.

Again, integration of ground-based activities to determine biospheric exchange will be of vital importance. From the ground-based work we will obtain the relationships between ecological and meteorological variables and the biosphere-atmosphere exchange fluxes of trace compounds. This knowledge may enable us to predict these fluxes on the basis of variables accessible by remote sensing. This will be a prerequisite for making estimates of global fluxes between biosphere and atmosphere. A key role will be played here by the development of automated instrument systems that incorporate technology derived from planetary missions. These systems will include automated gas chromatographs to measure concentrations and fluxes of important biogenic gases (e.g., NO, CH_4, and various sulfur gases), as well as sensors that determine quantitative chemical and biological information, such as soil microbial biomass, community structure, nutritional status, and metabolic activities. Other sensors will determine soil and sediment temperatures, pH, oxygen tension, soil moisture, and other parameters likely to affect microbial activity. The information collected by these automated systems will be relayed by satellite to data gathering and analysis centers. The incorporation of the biogeochemical processes deduced from in situ and remote-sensing measurements into appropriate models will enable us to forecast changes in atmospheric composition in response to biogeographic and climatic change.

Chemical cycling of elements within the atmosphere involves

photochemical reactions, processes involving exchange with aerosols and cloud droplets, reactions in these condensed phases, and a variety of scavenging or deposition processes. Again, the development of remote sensing techniques to measure the concentrations of key species in the troposphere, and their deployment on a variety of platforms are essential for testing our concepts of atmospheric cycling. Differential adsorption lidar (DIAL) is one such technique.

Biomass burning, especially in the tropical regions, has a very important influence on the chemical and physical characteristics of the atmosphere in the affected regions. Current efforts are under way to determine the characteristics of the emitted material and the relationship between the amount of biomass burnt and the amount of material emitted to the atmosphere. To estimate the regional and global emission rates as well as the biogeographical impact of burning agriculture, we will need remote sensing techniques to determine the size of the burnt areas and the amount of biomass that has been combusted. This could probably be done by existing or planned sensors, and is largely a problem of data processing and algorithm development.

The application of space science to the study of stratospheric chemistry, especially with regard to the stability of the ozone layer, which is essential for the protection of the biosphere from solar ultraviolet radiation, is treated in detail in the report of the Task Group on Earth Sciences.

GLOBAL ECOLOGY

In considerations of global biology, much discussion has centered (appropriately) on the biota as a system or series of interlinked systems whose collective metabolic activities influence the atmosphere, hydrosphere, and climate. The task group endorses the recommendations articulated in the Space Science Board Committee on Planetary Biology's (1986) report *Remote Sensing of the Biosphere* and the report of the Task Group on Earth Sciences calling for measurements of biome distribution and productivity, with special consideration of such critical problems as tropical deforestation, desertification, and the systems ecology of actual or potential agricultural regions. As has been pointed out in detail above, accurate data collection will necessarily involve coordinated measurements from observation platforms in space, high- and low-altitude aircraft, and from the ground. Advanced instrumentation systems will allow the integration of ground observations over appropriate areas and the gathering by satellites of data telemetered from remotely controlled, automated instrument packages. This will permit critical ecological data to be gathered from relatively inaccessible areas and make possible data acquisition on temporal and spatial scales that are unthinkable without space technology.

The task group further stresses that the recommended global biology research initiative presents significant opportunities for research in organismic ecology. Several types of information central to ecological thinking are poorly known from traditional ecological observation. One important question concerns the equilibrium of ecosystems. Are ecosystems in a steady state with regard to nutrient and energy flux, biomass, or taxonomic composition? If so, on what temporal and spatial scales does a dynamic equilibrium obtain? What are the effects of perturbations of varying frequency and intensity (such as storms, drought, flooding, fire, disease), and how does the ecosystem respond to such perturbations? Does it return to some approximation of the predisturbance state, and if so, on what time scale does recovery occur? What conditions of perturbation will "permanently" alter the ecosystem?

Models of community ecology (and related aspects of evolutionary biology) require information on the distribution of species populations within communities, as well as the spatial and seasonal distribution of phenological characters (flowering, fruit maturation, leaf fall, etc.). Such information can be difficult to obtain, particularly for tropical forests, where much of the Earth's diversity resides. Present satellite and aircraft remote sensors do not offer sufficiently fine-scale or discriminating data to resolve patterns of, for example, tree species distribution within a tropical forest. The task group recommends that technology be developed so that during the period 1995 to 2015 such resolution will become possible from aircraft. On a larger spatial scale, the task group recommends that technology be developed that will allow recognition of different successional stages within tropical forests. Such information, which is not available at present, is of theoretical importance and would be tremendously valuable to foresters charged with the management of forest resources.

Organismic ecological issues are of significant theoretical and practical value in themselves, but they are also relevant to the systems ecological measurements stressed in other reports. One simple question illustrates the point. Does the species composition of a community significantly influence its biomass and productivity, or will communities that have developed in physically comparable environments have comparable biomass and productivity regardless of taxonomic composition? The answer to this question will certainly influence our ability to integrate data and develop models for the evolution of the biosphere in geological time.

Much of the organismic ecological research community is only dimly aware of the potential that NASA's global biology program holds for ecological research. If that potential is to be realized during the time period in question, some program of mutual education should be established between NASA and this community prior to 1995. The task group recommends that a series of workshops be initiated in cooperation with the Ecological Society of America.

Global Ecology Models and Data Handling

Since our goal is to understand the mutual interactions of the biota with the geosphere, the hydrosphere, and the atmosphere, we need to integrate vast amounts of complex observational and experimental data. We need to formulate new generalizations, new theories. To achieve this goal, we will have to model, in both the short term and the long term, the dynamics of the surface of the Earth, since these biogeochemical processes simultaneously affect and affected by the biota. The task group emphasizes that modeling is uniquely important to global biology due to the very large number of parameters and the quantity of data relative to other physical and biological sciences.

Observational satellites, such as Landsat, or ground-based devices, such as those recording gas evolution, can measure numerous variables many times per day or even per minute. Experiments designed to measure seasonal fluctuations with high statistical validity may, in fact, contain evidence for circadian rhythms or other short-term phenomena. Such correlates would be lost by averaging in the course of data reduction, and would no longer be available when information is sought later that was not part of the original data reduction design. On the one hand, we cannot archive every bit of data for decades; on the other, we should not preclude serendipity by condensing data immediately to the logical needs of the proposed experiment. Each set of observations must be evaluated to strike a compromise. We need to develop an easy and rapid interactive system wherein the algorithms for data reduction can be altered as new phenomena are defined and the theories refined.

We will have to begin by modeling various components of the earth system. The many component submodels resulting from this effort serve two functions essential to science: they provide a framework for summarizing and evaluating the data, and they provide tests of our generalizations. However, the development of a comprehensive model that relates the diverse investigations will be one of the fundamental goals of global biogeochemistry in the twenty-first century. In this endeavor, NASA must provide leadership. No other organization has the physical resources, the intellectual talent (in-house and by consultation), and the global perspective to weld together this new scientific discipline. Of course, much of this leadership will take the form of designing special instruments and making new measurements. But NASA's most important role in global biogeochemistry will be to unite heretofore disparate disciplines around a common theme.

4
Controlled Ecological Life Support System (CELSS)

DEFINITION

To date, all manned space explorations have been relatively short, both in terms of overall time from start to completion of the missions, and also in terms of distance traveled from the Earth. Mission lengths have been limited by the carrying capacity of the spacecraft, the number of crew, and their life support requirements; all food and life-sustaining materials and supplies had to be carried in the craft when it was launched. When these consumables were nearly depleted, the mission space crew would return to Earth.

For long ventures in space, the resupply of life-sustaining materials from Earth is impractical, both technologically and in terms of cost. Extended Space Station missions, or longer term manned expeditions to the Moon or to Mars and beyond, will remain improbable until systems capable of regenerating life-sustaining materials (air, water, and food) become a reality. To this end, NASA is investigating the necessary scientific and technological requirements for a bioregenerative system, the Controlled Ecological Life Support System (CELSS).

RESEARCH OBJECTIVES

A two-pronged program is required to achieve the CELSS capability. First, the ability of plants and animals to grow, mature, and reproduce efficiently in the altered gravity of the spacecraft environment must be assessed. Through the normal photosynthetic mechanisms these same plants will serve in part as atmospheric purifiers. Whether we are concerned with life support in a spacecraft bound for the planets, or the establishment of a lunar base or martian colony, the basic principles are the same: we must know that biological systems will function productively under the environmental conditions at hand.

Second, the ability to cleanse and recycle the air and water needed to sustain human, plant, and animal life must be demonstrated. This requires engineering and development efforts of considerable magnitude. The engineering requirements include a plant growth chamber with appropriate light, humidity, and temperature controls; a dehumidifier to control excess moisture produced in the growth chamber by plant transpiration; a water purifier, to remove accumulations of toxic compounds; a food processing system to convert raw materials into edible material and convert, where possible, inedible matter into nutritionally usable matter; a waste processing system to recycle (chemically or biologically) human, plant, and animal waste into reusable materials; and an air purifier to remove toxic molecules from the atmosphere. The achievement of such a system clearly represents a difficult undertaking requiring long lead times.

Foremost among several important factors that must be considered in the design of CELSS is the rate at which organisms or physical-chemical devices produce or consume biomass, food, oxygen, carbon dioxide, potable water, and fixed nitrogen in response to variables such as temperature, light intensity, humidity, the nutrient medium used, and the composition of the atmospheric gas in which the organisms or devices operate. Automated sensing and data collection and interpretation are being emphasized in an effort to improve the efficiency, stability, and control of bioregenerative systems.

Specific research objectives in the CELSS development program include:

1. The development of practical methods for bioregeneration.

2. The determination of optimum environmental requirements of higher plants used in recycling systems.

3. Refinement of hydroponic and aeroponic plant growth techniques.

4. The investigation of lighting requirements.

5. Research into the use of algae as human food sources.

6. Definition of factors influencing algal productivity.

7. Inquiry into efficient biological waste processing methods.

8. Development of computer methods for operating and controlling bioregenerative systems.

ACCOMPLISHMENTS

Previous studies have indicated the approximate size, volume, and power requirements to accomplish water recycling, atmosphere regeneration, waste recycling, and plant growth sufficient to feed four to six humans on a space trip of several years' duration. This ranges from a minimum of 150 cubic feet to well over 200 cubic feet—the better part of a space station module.

Specific plants have been studied in the laboratory (for example, potatoes and wheat) for applicability to the CELSS environment. Optimum conditions for growth are understood and the utility of these plants as food sources and atmosphere regenerators is being evaluated.

A growth chamber capable of simulating the various environmental parameters in space (except microgravity and the complete radiation spectrum) and of the approximate size required is being designed to test the various systems needed for a working CELSS.

These research and development programs ultimately will lead to the selection of a set of plants and animals for use in a controlled chamber that could be included in a spacecraft or planetary or lunar habitat, and which would be capable of sustaining human life in extraterrestrial environments for very long periods of time.

The CELSS project is of great basic scientific interest as it involves research into large-scale, complex ecological systems involving humans. Also, it is of crucial operational importance for long-duration missions outside earth orbit. It is a program of formidable size and complexity from the management point of view. Nevertheless, it should be a project of high priority, both now and for the two decades beyond the Space Station era. Indeed, CELSS research could provide a connecting role—a focus of interest—for many of the scientific activities of the life sciences in NASA.

5
Space Biology

THE PROBLEMS

There are two basic questions to be asked in the area of space biology: (1) does the microgravity environment provide a research tool for studies on developmental and other biological phenomena; and (2) can plants and animals undergo normal development and life cycles in the microgravity environment?

Life has developed and evolved for at least 3.5 billion years in a relatively constant one-g environment. It is clear that both plants and animals have evolved mechanisms of detecting and responding to gravity. However, other issues are not as well understood: How do individual cells perceive gravity? What is the threshold of perception? How is the response to gravity mediated? Does gravity play a determinitive role in the early development and long-term evolution of the living organism? Studies of the early development and subsequent life cycles of representative samples of plants and animals in microgravity are of basic importance to the field of developmental biology. They are also essential to our ultimate ability to sustain humans for a year or more on the surface of extraterrestrial bodies or in spaceflight missions of long duration where resupply is not possible, and food must be produced in situ. More specifically, the areas important to study include:

1. Plant geotropic responses. The responses of plants to gravity and to light have been studied extensively. Roots grow down, shoots grow up. Spaceflight offers the unique opportunity to remove the gravitational stimulus from the developing plant system and to separate the gravitational input from other environmental stimuli known to influence plant growth (for example, phototropism and the circadian influences of the terrestrial environment). Spaceflight provides the opportunity to distinguish between the various tropic responses and to investigate the mechanisms of stimulus detection and response.

2. Animal systems, particularly embryonic systems (amphibian, fish, bird, mammalian) also have clear responses to gravity. For example, the amphibian egg orients itself with respect to gravity within a few minutes after fertilization, and during that short time the future of the embryo is established in the sense that the dorsal-ventral and anterior-posterior axes are established. Do we conclude therefore that the gravitational input is a required stimulus for the establishment of these axes? Clearly, the removal of gravity is a desirable, even necessary, step toward understanding.

The process of bone demineralization seen in humans and animals as a progressive phenomenon occurring during spaceflight is not only a serious medical problem, it raises the question of abnormalities in the development of bones, shells, and special crystals (such as the otoconia of the inner ear) in animals developing in microgravity. The study of such abnormalities should provide insight into the process of biomineralization and demineralization.

3. In addition to the scientific need to study basic plant and animal interactions with gravity, there is a practical need to study the responses of animals, and particularly plants, to the "normal" environment expected on future spacecraft. The long-term presence of humans in space (either in interplanetary flight or in an extraterrestrial station) is an integral part of planning for future space explorations (for example, manned Mars missions or bases). When the point is reached where it is no longer cost effective or logistically possible to resupply the spacecraft or habitat with water, atmosphere, and food, ways must be found to recycle all these components in a controlled ecological life support system (CELSS). Experiments in spacecraft aimed at determining which plants and animals are most efficient and best suited for such a system are badly needed. While much work can be and is being done on the ground, flight work is essential. Again, however, the

experiments needed require long-term (many months) spaceflight. Ideally, this spaceflight would be of sufficient duration to demonstrate that several generations of plants can flourish (for instance, can soybeans germinate, grow normally, produce an optimum crop of new soybeans for food and new seed for ensuing crops?). All of this biological cycling, plus the development of water and atmosphere recycling equipment, the reclamation of wastes, etc., requires a long-term engineering enterprise.

WORK TO DATE

Although both plants and animals have been flown on space missions, an adequate data base has not yet been obtained. There are several reasons for this. First, the opportunities for flight have been rare. Second, the environment provided for living material in spaceflight has not yet been optimized. For example:

- The threshold of gravity sensing systems in plants and animals is unknown, but there are reasons for believing that, in certain organisms, it is in the range of 10^{-5} to 10^{-6} g. It is calculated that, at these gravity levels, sedimentation and thermal convection are no longer operative. These two physical factors are believed to be the principal intracellular phenomena in effecting the gravitational input within the cell—a testable hypothesis at the appropriate microgravity level. Spacecraft to date have not provided that gravity level. Most flights have been in the range of 10^{-3} to 10^{-4} g.
- In the study of embryonic material in particular, most experiments have of necessity been done with eggs that were fertilized on the ground, well before orbital flight, so that the critical g-sensitive time period immediately after fertilization was spent at one-g.
- Most U.S. flights have been too short in duration (a few days) for many observations to be made.

Early flights (both U.S. and Russian) have amply demonstrated that the flight of living material is possible. The microgravity environment encountered to date (10^{-3} to 10^{-4} g) has not been low enough or long enough to allow us to answer the question of threshold and response mechanisms. Cells seem to develop normally, although growth patterns (particularly of plants) are adversely affected even at these levels. Thus, while considerable experience has been gained and will continue to be gained in orbital flights of days to weeks in duration, much lower gravity levels (10^{-5} to 10^{-6} g) and much longer flight duration (months) are needed.

FUTURE WORK

Continuing flight of the Space Shuttle will provide limited opportunity to expose living systems to microgravity, but it appears that the opportunity to do the critical experiments must wait for the development of spacecraft capable of providing both lower gravity levels and flight times of months to years. The first such capability may well arise with the advent of the Space Station—in 1995 and beyond. Whether such an environment is provided by a space station, free flyer, or platform, the real requirement for this work is an environment that is well defined in its physical parameters—that is, long-term, uninterrupted low g. A space station module, if mounted on the center of mass of the space station and not perturbed by human or machine activity, could provide a minimal environment—10^{-5} g. Otherwise, a free flyer or platform may be required. Only flight opportunities such as these will permit a rigorous answer to the problems outlined above.

Special Requirements

The special requirements needed to ensure successful space biology experiments in a microgravity environment can be listed as follows:

1. A "quiet" laboratory or platform that has an environment of 10^{-5} to 10^{-6} g at <1 Hz is needed for this work. This requirement is shared by many materials science activities as well.
2. This environment must remain undisturbed by human or machine activity for long periods of time (many months). Special "damping" procedures may be needed.
3. Periodic sampling is needed, but it should be done in such a way as to leave the remaining material undisturbed. Sampling must obviously be carefully planned and minimized to preclude vibrations and other unwanted gravitational forces.
4. Appropriate incubators and growth chambers are needed for cells, simple organisms, plants, and animals.

5. In order to investigate the problem of gravity thresholds in living systems, an on-board centrifuge is needed. This centrifuge should provide a one-g control for microgravity experiments, as well as the capability to explore a range of gravities between 10^{-5} and one-g, in order to study CELSS candidates and gravity thresholds for certain phenomena.

6
Human Biology and Space Medicine

INTRODUCTION

The Space Science Board's recent report, *A Strategy for Space Biology and Medical Science for the 1980s and 1990s* (National Academy Press, 1987), proposes a broad program of biological, physiological, and psychological research for NASA into the next century. The report concludes that the field of space biology and medicine is in its infancy and emphasizes that the success of any long-term manned program is based upon the support of a well-balanced, vigorous program of research over the next 10 to 20 years. The task group supports these conclusions and notes that its own recommendations complement and extend those themes to the year 2015.

This task group wishes to emphasize several of the conclusions contained in the Space Science Board report:

- Basic and clinical research complement one another in that clinical observations often pose basic research problems. Conversely, solutions to medical problems usually depend on an understanding of the underlying biology. The design, execution, and interpretations of these experiments should honor this complementarity.

- Under the best of circumstances, opportunities for flight experiments will be infrequent and expensive. Extensive ground-based simulations and experiments must precede space experiments whenever possible.
- Most of the experiments will require manned missions both because humans are frequently the subject of choice and because many experiments require observations and manipulations too complex to be performed remotely. We will be establishing the ranges of human physiological and behavioral responses to incrementally increased exposure to microgravity. All crew members should contribute to this data base without compromising their confidentiality, welfare, or performance. There must also be experiments on plants and animals. We should identify species suitable for the range of experiments and optimize our holding facilities for them. Ground studies should select desired mutants, explore their physiologies, and map and sequence relevant genes.
- In order to interpret microgravity experiments, we must have adequate controls. Usually, these will involve parallel experiments performed at one-g in space, duplicating all of the biological and physical variables of the microgravity experiment. Building the required centrifuge is a major engineering challenge.
- If the United States is to have the option of manned spaceflight of several years' duration, we must define and ameliorate several potentially harmful effects of spaceflight. Beyond this, we must determine whether space provides opportunities better than those existing on Earth to study biology and medicine. Our past flight opportunities have been so restricted that we are not well prepared to project research strategies to 2015.
- Men and women may be called upon to perform a broad range of jobs in construction, observation, and manipulation while in planetary orbit or interplanetary flight. We need to research the optimal interactions of the crew while working with under stressful conditions, while working with one another, and while working with a complex array of equipment and computers.

EXPERIMENTAL USE OF ANIMALS

Areas of special concern include animal use. There has been little substantial progress made in the development of animal holding units and research modules for the U.S. and Soviet space programs. The considered use of animals in space research is necessary for an orderly, thorough program. In order to qualify humans for long-term flights or for repeated short-term flights, our scientific data base must be greatly extended.

The uncertainties and potential dangers (from weightlessness, radiation, and so on) of long-term space missions underscore the need for carefully prepared animal flights in the period leading up to those missions. Gaining the experience and knowledge we need from animal research will demand a clearly enunciated commitment by NASA and the assignment of adequate priorities and resources.

Approach to Scientific Questions

Planning the research and development program for a space vehicle must incorporate factors considered in several sections of this and other chapters. A principal requirement for the advancement of the scientific objectives related to the vehicle is for a well-planned, carefully constructed, multipurpose, residential, ground-based vehicle simulator for intermediate and long-term investigations. While this would not permit study of critically important stresses (such as radiation and weightlessness), it would allow assessment of the structural features of the vehicle and the characteristics of their outgassing products; the utility and suitability of the arrangement of equipment; and the unique requirements for special research laboratories and animal holding and research facilities. Such a facility would also allow testing of modular design features and would be suited for long-term studies of nutrition, health care problems, and psychosocial interactions. Such design characteristics should be evaluated on short missions prior to a commitment to longer-term space missions.

NEUROSENSORY PHYSIOLOGY

Introduction

Since its inception, NASA's neurosensory physiology research program has centered on the neurovestibular area largely because of the space motion sickness (SMS) problem that surfaced in the Apollo program. This focus is both appropriate and justifiable in view of the continued SMS incidence affecting over 50 percent of the crew on Shuttle flights, and in view of the resulting decreased crew efficiency, as well as the potential impact on early extravehicular activity (EVA) requirements, and the potential problems for SMS-affected crew members during arrival and transfer at the Space Station.

The goal of understanding the neural mechanisms of adaptation to microgravity is an enormous task requiring a broad-based program of investigation. Devising predictive SMS tests on Earth and rational therapies for the prevention or amelioration of SMS clearly rests on an intensive research effort aimed at understanding the mechanisms involved in human habituation to microgravity and readaptation on return to Earth.

This broad-based neurosensory research program must emphasize a systematic, well-coordinated series of inflight experiments involving human and subhuman surrogates as research subjects. A relatively comprehensive ground-based research program is currently in place; however, a firm commitment for an extensive inflight program is mandatory if progress in this complex area is to be forthcoming. In this context, the validation of a suitable animal model for invasive experimentation would accelerate research progress.

Extensive inflight experimentation will be required for several reasons:

1. SMS and other vestibular phenomena are unique to space (microgravity environment).

2. Microgravity cannot be simulated on Earth for sufficiently long time intervals (30 s in KC-135 parabolic flights).

3. SMS provocative stimuli (microgravity and head/body movements) are different from causative stimuli in motion sickness of one-g. Such movements at one-g do not result in motion sickness.

4. One-g motion sickness susceptibility tests do not predict susceptibility aloft. The correlation coefficient is low.

5. Habituation exercises such as gymnastics and acrobatics may have a salutary effect on performance in other one-g motion environments, but there is apparently no transfer to the microgravity environment, and such techniques have not been demonstrated to be effective in preventing or mitigating SMS symptoms.

Very probably no "magic bullet" will be found to prevent SMS. In this case NASA's research strategy will have to rely on a coordinated, systematic approach to understanding the complex operant neurosensory mechanisms. Even if the "magic bullet" for SMS were found, the requirement to do extensive neurophysiological (vestibular) research in space would still exist. The adaptations of this system are of great intrinsic interest. We have evolved on planet Earth for eons in a one-g environment. Only in microgravity can we remove this gravity vector and physiologically "dissect" the sensory nervous system in a nondestructive mode. For this reason alone, and for the immense benefits that will accrue to neurophysiological and medical knowledge and understanding of these complex phenomena, the systematic investigation of the vestibular/neurosensory system is a *scientific imperative* for the Space Station.

Background

Flight Experience

For many reasons, progress in human physiological research in space has been limited. Experimental sample size has been and will continue to be small. In addition, experiments generally require at least two crew members (experimenter and subject). Finally, there has been a long hiatus in flight opportunities between Skylab and Shuttle programs. The dearth of flight research opportunities reflects the low priority given to life sciences research in general.

The first dedicated Life Sciences Mission on the Shuttle (SLS-1) will carry experiments proposed in 1978. This is to be followed by SLS-2, carrying animal experiments also proposed in 1978. Data in the neurovestibular area have been measured by the European Space Agency (ESA) mission flown on the Shuttle, Spacelab 1 (SL-1), the German D-1 mission, detailed supplementary objectives (DSOs or mid-deck measurements), and limited preflight and postflight testing. These results will be referred to in the sections to follow. There is no current provision for a vestibular research facility on board the Shuttle, although time and space will be made available for single SMS/neurosensory experiments as these mature and are ready for flight.

Given the formidable obstacles to inflight life sciences research outlined above, both reason and past experience indicate that the bulk of any proposed systematic, well-structured scientific research program in the neurophysiological discipline will have to await the Space Station era. All available Shuttle flight opportunities will be utilized in the interim. These will be helpful, but inadequate to support the research program currently envisioned.

Improved technologies for invasive and noninvasive experimental methods will doubtless be developed in the coming years. Some of these will lend themselves to relatively easy incorporation into the inflight investigative program; some will not. However, much vital information can be obtained using present state-of-the-art technology. Special facilities will be required for the execution of the research program detailed in this chapter; some of these will influence the design of the Life Science Research Module. These include:

1. A "space sled" (linear body accelerator) with triaxial seat capability.
2. A subject rotator (also with triaxial capability).
3. A multipurpose, variable-g centrifuge to accommodate animals, plants, and eventually humans.

The latter facility would be of immense value in several disciplines. In neurosensory research it will permit observation and quantification of sensorimotor adaptation at various g levels during prolonged stays in microgravity. It would be invaluable in determining the optimal g level to prevent or reduce the adverse effects of microgravity such as cardiovascular deconditioning, bone demineralization, muscle atrophy, and abnormal plant growth.

Research involving subhuman surrogates is detailed in *A Strategy for Space Biology and Medical Science for the 1980s and 1990s.*

Space Motion Sickness

As previously mentioned, neurosensory research efforts in the past have focused on space motion sickness (SMS). This is a special form of motion sickness that is experienced by some individuals during the first several days of exposure to microgravity. The syndrome may include such symptoms as depressed appetite, a nonspecific malaise, lethargy, gastrointestional discomfort, nausea, and vomiting. As in other forms of motion sickness, the syndrome may induce an inhibition of self-motivation, which can result in decreased ability to perform demanding tasks in those persons who are most severely affected. Table 6.1 gives the symptom incidence in 72 astronauts. The syndrome is self-limited.

TABLE 6.1 Space Motion Sickness Incidence and Symptom Complex

Symptom Incidence	Number	Percent
Vomiting	31	86
Anorexia	28	78
Headache	23	64
Stomach awareness	22	61
Malaise	21	58
Lethargy	19	53
Nausea	18	50
Drowsiness	8	22
Disequilibrium	7	19

Overall SMS Incidence: 36 of 72 astronauts (50%)

Complete recovery from major symptoms—in other words, adaptation to the spaceflight environment—occurs within two to four days. After complete adaptation occurs, crew members appear to be immune to the development of further symptoms, although the Russians have reported repeat episodes. This development of immunity to further SMS symptoms was eloquently demonstrated by rotating chair tests, designed to provoke an SMS response, that were conducted inflight during Skylab missions.

The etiology of SMS is currently under intensive investigation. There are two main hypotheses advanced to explain SMS: Sensory or neural mismatch (a variation of the sensory conflict theory) and the fluid shift model. Research data currently lend more credence to the former theory.

Prediction of susceptibility has been an objective of the SMS research. Various approaches ranging from the use of questionnaires, psychodynamics or personality traits, vestibular function tests, physiological correlates, and tests in specific nauseagenic environments have been directed toward the question of SMS susceptibility. The correlations between the selected predictors and motion sickness have been of limited use in predicting susceptibility. Individual variations in preflight experience, medications, inflight tasks (i.e., mobility), and personal strategies for symptom management have further compounded the problem.

Vestibulo-Spinal Reflexes

Two of the more dramatic responses to orbital flight have

been postural disturbances and modified reflex activity in the major weight-bearing muscles. Monitoring the Hoffman A reflex, which takes advantage of the powerful and established anatomical pathways that link the otoliths and spinal motoneurons, has been selected as a method of monosynaptic spinal reflex testing when performed in conjunction with linear acceleration to assess otolith-induced changes in one group of major postural muscles (soleus). Second, extensive dynamic postural testing with a moving platform was performed before and after the flight of Spacelab 1. The Hoffman reflex amplitude, as reflected by the otolith-modulated motoneuron sensitivity, was low in flight after neurovestibular adaptation to spaceflight, and its postflight potentiation may have been dependent on rate of adaptation. The degree of inflight SMS symptoms was related to preflight and postflight Hoffman reflex amplitude. Dynamic posture tests revealed significant deviations from the results obtained before flight. The strategy used by the individuals for balance on the moving platform was modified, and their behavior indicated a decrease in awareness of the direction and magnitude of the motion.

Proposed Research

Overview

Investigations using noninvasive methods and human subjects during long-duration orbital flight will be required for an understanding of human adaptability to microgravity and the return to Earth. Possible areas of investigation include:

1. Prediction of susceptibility.
2. Profiles of adaptability.
3. Training countermeasures.
4. Pharmacological countermeasures.
5. Physiological variables.
6. Neural mechanisms of SMS and adaptation to microgravity.
7. Vestibular function tests.
8. Vestibulo-spinal reflex mechanisms.

Visual System

The human visual system is addressed in the body of this

discussion principally in the context of its relationship to the vestibular system. Vision may compensate in large measure for modified otolith sensitivity. It helps in spatial orientation, and is essential to motor coordination. Other aspects of vision and the visual system must also be addressed.

Visual acuity, contrast discrimination, and radiation effects must be investigated, particularly on long-duration missions. The former two functions have been studied on short-term Shuttle flights, and no significant decrements have been found. The effects of long-duration flights on these functions are unknown and must be determined.

Visual "light flashes" were observed by Apollo astronauts and were more systematically studied on the Skylab 4 mission. During most of the Skylab 4 mission, these flashes averaged 20 per hour; however, flashes increased to 157 per hour when the Skylab orbit passed over the center of the South Atlantic anomaly. The flashes are believed to be due to high-energy heavy particles (cosmic rays) and have been reproduced in humans in the laboratory by exposure to high-energy ionizing particles at the Berkeley Bevalac facility. The flux of these particles may be expected to increase during any missions beyond the Earth's magnetosphere as well as during polar and high-inclination earth orbits. Thus, we must assess the effects of this radiation on nondividing cells of the retina and central nervous system before long-duration manned missions beyond Earth's magnetosphere are attempted. The radiation hazards to humans in space are addressed in greater detail in the "Radiation Effects" section of this chapter.

Tactile and Proprioceptive Systems

The gravity vector is a fundamental factor in human spatial orientation, which results from the integration of a complex of sensory inputs including visual, vestibular, tactile, and proprioceptive. The latter two systems are important both in spatial orientation and in postural control.

In some individuals static visual cues become increasingly dominant in establishing spatial orientation in microgravity. Other subjects are more "body oriented" and align their exocentric vertical to be along their longitudinal body axis, and perceive the body axis relative to placement. Such individuals exhibit no problems in spatial orientation aloft even in the absence of visual cues for vertical orientation. Further, these individuals appear able to strengthen their perception of subjective verticality by using localized tactile cues, especially by pressure exerted on the soles of their feet. It is evident that the nature of proprioceptive changes in microgravity must be addressed in detail, particularly neck and joint angle sensors and the role of localized tactile cues in the perception of body verticality.

Perceptual and physiological data from Spacelab 1 underscore the need for additional systematic study. Early in a given spaceflight (before adaptation), sudden drops were perceived as falls or drops on Earth—linear translations—and felt much as they did preflight. Hoffman reflex changes at this time were similar to those observed preflight. Later, inflight drops were perceived as linear translations that were sudden, fast, and hard, with the crew not being aware of the position or location of their legs and feet and the absence of a falling sensation. They also exhibited difficulties in maintaining "balance" following the drop. In these late inflight drops, the Hoffman reflex was not potentiated. Finally, postflight drops were perceived as identical with the late inflight drops, namely, lack of awareness of position and location of feet. Again, the drops were perceived as sudden—not a falling sensation, but rather a feeling that "the floor came up to meet them."

Experimental programs concerning spatial orientation must therefore identify and quantify the sensorimotor adaptation in microgravity as well as in the postflight "carryover" and readaptation to one-g periods. Such experiments will provide information regarding the psycho-physiological basis for establishing spatial orientation aloft.

Postural mechanisms require investigation in long-duration space missions. Postural activity is the complex result of integrated orientation and motion information from visual, vestibular, and somesthetic sensory inputs. These inputs collectively contribute to a sense of body orientation and, additionally, coordinate body muscle activities that are largely automatic and independent of conscious perception and voluntary control. Changes in the adaptation of automatic postural systems to microgravity have not been studied systematically, but impairment of voluntary pointing accuracy and misperception of static limb and body position have been noted, as have postflight impairment in walking, standing (eyes closed), and negotiating rapid turns. Microgravity is the only external environment condition in which segments of previous one-g postural control are no longer relevant and in which vestibular inputs are altered on a long-term basis. Microgravity affords a unique opportunity to investigate the adaptation of automatic postural control systems and to augment our understanding of adaptive mechanisms involved in the voluntary muscular system.

BONE AND MINERAL METABOLISM

Introduction

Loss of calcium and phosphate from bone (osteopenia) continues at about 0.4 percent of total existing bone mass per month. The effect is especially marked in the weight-bearing bones of the legs and spine. There is no indication that this osteopenia abates with longer flights. The resorbed mineral may affect various organs, especially the kidneys. Bones could fracture under the extreme stress of heavy work or upon return to one-g.

This poorly understood phenomenon is one of the major dangers posed by flights of several years. There are no immediate prospects for countermeasures based on exercise, diets, or drugs. Although empirical procedures may be found, they cannot be assured. The task group believes that a major research investment is mandatory before serious planning begins for spaceflights of several years' duration.

Background

Flight Experience

A variety of studies of humans during long-term bed rest, of humans in space, and of rats in space have shown that prolonged inactivity and weightlessness result both in significant and continuing losses of calcium from the skeleton and nitrogen from muscle, and in notable atrophy of both body systems. These changes were consistent but quite different in degree from subject to subject. In the longest bed rest studies (7 months) and in the longest orbital spaceflight during which metabolic measurements were made (3 months), the rate of calcium loss was as great at the end of the studies as it was soon after the start. In the severe paralysis of poliomyelitis, calcium losses led to x-ray visible osteoporosis in the bones of the lower extremities as early as 3 months after paralysis. While the overall rate of calcium loss in Skylab astronauts was 0.4 percent of total body calcium per month, the loss was estimated to be 10 times greater in the lower extremities than in the rest of the body (based on bed rest studies of calcium losses by metabolic balance compared with decrease in bone calcium density). This could lead in 8 months of flight to a decrease in bone density in the legs similar to that noted in paralytic poliomyelitis. In longer flights, if mineral loss were to continue at a similar rate the bones of the legs might fracture during physical work in as little as 9 to 12 months, especially at gravities approaching one-g. Studies of immobilized rabbits showed marked decrease in strength of tendons and ligaments after only 1 month. Thus, strains, sprains, and even ligament tears may be more likely to occur, and at an earlier time than bone fractures.

Cellular Mechanisms

The cellular mechanisms of mineral loss are unknown. Excess excretion of calcium associated with increased hydroxyproline in the urine in humans is indicative of increased bone resorption. Histologic examination of the bones of the rats on Cosmos showed suppressed bone formation; it is difficult, however, to apply these results directly to humans because of differences in rat bone physiology.

In more recent research, bed rest studies under NASA sponsorship have been continued in search of so-called "countermeasures" that could be applied to astronauts in space to suppress or prevent calcium loss. All of the mechanical procedures tested thus far have been ineffective. Correlative observations have indicated that one would have to devise some procedure for use in flight that would provide the equivalent force on the skeleton of 4 hours of walking per day. A high calcium and phosphorus diet reduced calcium loss for up to 90 days only. Some promise has been noted in certain of the diphosphonates, compounds that bind to bone crystal and tend to inhibit bone resorption. These countermeasure studies are being continued.

At the same time, with support from the National Institutes of Health, a variety of studies are being conducted on the basic mechanisms of the effects of mechanical forces on bone dynamics and development. Such studies may give insight into the bone loss problem in space. Conversely, development of effective countermeasures to bone loss in space may contribute to improved therapy or management of osteoporosis, which is characterized by gradually decreasing bone mass and strength, and is the most prevalent clinical disorder of bone.

Possibility of Urinary Tract Stone Formation

The hypercalciuria associated with loss of mineral from bone in spaceflight might increase the potential for stone formation in the urinary tract. Although 75 to 80 percent of renal stones contain calcium, the likelihood of stone formation will depend not only on increased urinary concentration of calcium, but also on other factors such as urinary pH, concentration of inorganic elements (magnesium, potassium, and phosphorus), and concentrations of organic compounds (uric acid, citrate, and oxalate). Bed rest studies have shown a slight rise in urinary pH and a lack of change in urinary citrate, which in ambulatory states rises with increases in urinary calcium. Both of these factors, if also noted in spaceflight, would favor decreased solubility of calcium salts. These considerations suggest that research ought to be continued on urinary tract stone formation in relation to microgravity as a significant possibility during long spaceflight. The likelihood of such an occurrence may be small, especially if care is taken to maintain abundant urine volumes; nevertheless, such stone formation might be catastrophic to health and function for the astronaut involved, and thus to success of the particular flight.

Proposed Research

Characterization

Over at least the next 10 years, research in bone physiology and metabolism related to space is likely to be involved primarily in the search for effective ways of protecting the skeleton from the decreases in mass associated with weightlessness and diminished physical activity. The initial phase will be a continuation of human bed rest studies and immobilization or suspension studies in primates and rats, respectively, as the only practical models for weightlessness. The principal "countermeasure" studies thus far have been with humans subjected to bed rest. Thus far, in these studies no physical procedure tested has been helpful in preventing disuse osteoporosis. Among the biochemical modalities, trials of only two different diphosphonate compounds (EHDP and di-chloro) have suggested potential for usefulness. The rat and primate studies, including a number of rat studies on U.S.S.R.-Cosmos flights, have been mainly observational, but with the intent of obtaining data on the rate, degree, location, and pathophysiological processes of the bone loss occurring in inactivity and in weightlessness.

While appropriate plans for Space Station research relative to the musculoskeletal system are being addressed, certain studies will be especially pertinent to learning more about the effects of long-term weightlessness, on the musculoskeletal system. The most obvious animal study to conduct in Space Station would be exposure of successive groups of rats to 30, 60, 90, or more days in weightlessness, followed by various biochemical, radiological, and histologic examinations of muscle and bone to determine the rate and course of bone loss and to obtain further insight into its basic mechanism. The rat is not the ideal model for such studies, however, because of its ever-growing skeleton. Plans should be made, therefore, to carry out similar studies on animals with skeletons that mature. These might include cats, dogs, small pigs, and—perhaps at a later date—primates. The task group again emphasizes the desirability of focusing research on a limited number of well-characterized species.

In line with the NASA principle of extending knowledge of the effects of spaceflight by progressive extension of the duration of flight, the answer to the question of whether calcium loss will continue beyond some critical level of bone density needs to be obtained by studies of astronauts in longer exposure to microgravity. Payload specialists should be studied in space to at least 6 months, and depending on results, possibly on to 9 to 12 months. Such studies should include not just measurements of preflight and postflight bone density. Rather, they should include metabolic studies of certain individuals, analyzing dietary controls and total excreta. This would allow a rather precise assessment of the pattern and extent of continuing bone and muscle loss. The need for imaging techniques required to facilitate this research is discussed in Chapter 7. It is imperative that we progress with this research before we commit to spaceflights of several years, as required for Mars exploration.

Cellular Mechanisms

The current models in bone research are bed rest for humans, tail suspension for rats, and immobilization for primates. An effort should be made to develop other animal models that are less expensive and more manageable than primates, and have a more appropriate skeletal system than rats. As for experimental techniques, relatively little has been done thus far among space physiology investigators in the area of kinetic studies for assessment of rates of bone formation and resorption and the effects thereon of various procedures or agents. Radioactive isotopes can be used in animal studies at proper dosages. Kinetic studies of humans in space will be more difficult. Use of stable isotopes is increasing generally despite their high cost, and presumably by 1995 appropriate studies can be designed for humans in microgravity. Some medical bone physiologists are increasingly interested in biophysical and biomechanical influences on bone (Kroc Conference in 1983 on "Functional Adaptation in Bone Tissue").

Development of bone cell and tissue cultures will also be required for progress in understanding bone and mineral metabolism in a microgravity environment. As they mimic hormonal and mechanical stresses, their various compositions and characteristics can be manipulated more readily than can intact animals. More precise measurements of magnetic and streaming potentials and of piezoelectric energies are becoming available for use in a wide variety of studies, including the interrelationships of mechanical and electrical forces and biochemical signals in bone cell cultures. Despite the steady development of more sophisticated techniques, the capabilities for diet control and specimen collections of metabolic balance studies will still be needed aboard space vehicles. Whatever the important insights we gain from organ, tissue, and cellular studies into mechanisms and how to influence them, there will still be the need to measure periodically changes in rates, masses, and patterns of flow of key elements in the whole animal or whole human being.

Nutrition

Numerous previous studies unrelated to space have indicated that increasing the protein intake increases the urinary excretion of calcium. Therefore, the level of protein in the diets of astronauts, hitherto selected by them as quite high, needs to be reconsidered for its possible relationship to potential for urinary tract stone formation and to the rate of loss of mineral from the skeleton. At the same time, some degree of uncertainty exists as to whether or not the high phosphate content of meat is partially protective. The Space Station would be the logical clinical laboratory in which to settle the question of the best calcium, phosphate, and protein content and proportions in the diets of astronauts subject to months and years of weightlessness.

The possible benefit of nutritional manipulation has been tried among various efforts to find a means of protecting the skeleton in weightlessness. In long bed rest studies sponsored by NASA (V. Schneider and Associates), high calcium and phosphate intakes kept calcium balance from going negative for the first three months of bed rest, but during the fourth month increasing fecal calcium excretion resulted in continuing losses of calcium (negative balance). A similar study of the effects of high calcium intake might be repeated in the Space Station, but it probably would not have high priority.

Summary

Review of the current state of knowledge regarding bone demineralization and the possibility of urinary calculi formation in space indicates that considerably more research will be needed in this area during the period 1995 to 2015. This well-known constellation of problems will require, at the least, careful monitoring during prolonged stays in space, even if countermeasures that are efficacious in missions of shorter duration have been developed. Bone demineralization is probably the most significant hazard for long residence in space at present. A fully effective countermeasure, or indeed a full understanding of the involved mechanisms, will not be available by 1995 and will require a major investment in understanding the fundamental biology of bone development, formation, and resorption.

MUSCLE METABOLISM

Introduction

After a few days of exposure to microgravity, the urinary excretion of nitrogen compounds increases and muscle atrophy begins. These effects may compromise the ability of astronauts to do their jobs. They may not be able to withstand the stress of one-g upon return to Earth; the continued excretion of nitrogen may have deleterious hormonal and nutritional effects. Although exercise, diet, or drugs may ameliorate these effects, the task group anticipates that a successful prophylaxis will be based on a fundamental understanding of the control of the expression of the genes encoding the structural and regulatory proteins of muscle.

Background

Flight Experience

The increased urinary excretion of nitrogen by astronauts in Skylab reflected mainly muscle loss as is observed in bed rest, but was variable and generally greater in degree. At least a small part of this excess nitrogen excretion could reflect gluconeogenesis related to observed increased cortisol release, although the increases in plasma cortisol in both Skylab and a recent bed rest study were quite modest. The possibility of some other influence similar to the "toxic" factor of severe illness cannot be ruled out. However, most of the atrophy occurs in antigravity muscles, which are no longer load-bearing. Of these various possible factors contributing to the excess excretion of nitrogen, muscle atrophy is clearly the main one.

In all nine Skylab astronauts, the high level of nitrogen excretion continued unabated for the duration of flight (up to 84 days). This indicates a serious malfunction not likely to reach a new steady state until an extreme degree of atrophy is reached. This nitrogen loss was accompanied by losses of 15 to 30 percent of muscle mass and strength in the lower extremities. This poses a significant handicap to vigorous work in the gravity of Mars or on return to Earth. The considerable and time-consuming exercise activity of the astronauts on Skylab 4 resulted in somewhat lesser losses of muscle mass and strength than on the earlier flights, but were obviously not adequate to be fully protective.

In spaceflight studies of rats in the U.S.S.R.-Cosmos series of flights, there was a differential atrophy of the various types of muscle fibers and many focal histopathologic changes with random deletion of myofibrillar filaments. There was a loss of muscle force and elasticity and some specific changes in enzyme activity. In the rats in these Cosmos studies that were subjected to centrifugation, these muscle changes were largely prevented.

Ground-Based Studies

Bearing directly on this human problem are NASA-sponsored animal studies of muscle atrophy due to disuse. These studies emphasize efforts to determine the physiological and biochemical mechanisms underlying muscle atrophy. They also are directed toward development of noninvasive methods of measuring muscle mass and toward searching for useful countermeasures. Although the mechanism of the process of atrophy remains unknown, certain aspects have become evident. Muscle atrophy is accompanied by decreased synthesis of muscle protein and by some degree of increased degradation. As shown in rats that are suspended (hind limb unloaded), loading and stretching of otherwise inactive leg muscles prevented muscle atrophy and stimulated protein synthesis; the addition of electrical stimulation increased protein synthesis markedly. As shown in muscle cell cultures, stretching stimulates protein syntheses.

The uncertain value of physical exercise for suppressing muscle atrophy in human flight has been noted previously; no controlled studies of exercise in flight have been attempted.

Proposed Research

NASA-sponsored research is now addressing basic mechanisms relating muscle demand or load to hypertrophy, and decreased demand to atrophy. What is the signal and sequence of biochemical steps for initiating increased protein synthesis and deposition in muscle filaments, and what communicates a message to slow down protein synthesis? Answers to these questions would have an impact on muscle research far beyond spaceflight.

Using various animal and human studies, we need to determine which muscles and fiber types are primarily affected in relation to the duration of exposure to microgravity. The relationship of muscle fatigue to microgravity exposure and the usefulness of various exercise regimens in ameliorating the effects of microgravity need to be studied. What microanatomical, enzymic, and biochemical changes occur in myofilaments, in the muscle tissue of the walls of the veins, in myotendinous junctions, and in tendons and ligaments when subjected to microgravity?

A variety of techniques are available for muscle research: electron microscopy, electromyography, CT scanning, and stable isotope metabolic studies. The effects of electrical stimulation of muscle have begun to be studied, but the possible combinations of frequency, voltage, and current are almost without limit. These are but a few of the permutations possible in undertaking to increase our knowledge of muscle physiology and biochemistry as influenced by microgravity. These existing technologies should be coupled with developing techniques in immunochemistry and in recombinant DNA and gene cloning.

In order to understand changes of muscle mass and strength, we must understand their underlying cellular and molecular mechanisms. The genes encoding many major proteins of muscle, as well as their controlling elements, have been sequenced. Our goal is to relate mechanical stress, hormonal levels, and nutrition to the control of expression of these genes.

CARDIOVASCULAR, PULMONARY, AND RENAL SYSTEMS

Introduction

Investigation of cardiovascular and pulmonary physiology in manned spaceflight will continue to be an important endeavor well beyond 1995. Clearly, the cardiovascular, pulmonary, and renal systems are crucial to health. At this writing, human cardiopulmonary and renal response to short-term exposure to microgravity seems to have been relatively free of major threats to well being and performance in spaceflight. However, we now have only a rudimentary glimpse of the total picture of optimal preparation for spaceflight, physiologic behavior during exposure to microgravity, and reaccommodation to a one-g environment.

Background

The cardiopulmonary and renal systems readily adapt to microgravity. Adaptations to relatively short-duration flights (7 to 10 days) are quickly reversed upon return to the one-g environment. It now seems likely that flights of 28 to 237 days, the shortest Skylab and longest Salyut missions, respectively, result in somewhat more extensive although qualitatively similar adaptations, but demand a proportionately greater time for readaptation to one-g. Two examples of postflight problems are orthostatic intolerance and diminished exercise capacity.

Orthostatic intolerance is characterized by a cluster of symptoms that follow standing: lightheadedness, resting tachycardia, labile blood pressure, narrowed pulse pressure, and presyncope or syncope (fainting). Diminished exercise capacity is the observed decrement in ability to perform given amounts of work and is usually measured by duration of treadmill or stationary bicycle exercise up to a maximum level of oxygen consumption. At current levels of experience, both orthostatic intolerance and diminished exercise capacity become more severe with longer exposure to microgravity and require more lengthy recovery times after returning to Earth. There appears to be no qualitative difference between short- and medium-term microgravity exposure with regard to cardiovascular well being and performance in space. For the Shuttle era, researchers remain concerned with devising and refining countermeasures to prevent or palliate cardiovascular problems associated with the return from microgravity to earth gravity (one-g). Current candidates for countermeasures include preflight and inflight exercise, application of lower body negative pressure (LBNP) during spaceflight, fluid loading prior to reentry, and rehabilitation after return to Earth.

Low blood volume, fluid shifts, orthostatic and LBNP intolerance, reduced exercise capacity, and decreased heart size have all been observed after exposure to microgravity for up to 6 months. However, the effects of exposure to microgravity beyond 9 months are entirely unknown. This is of great concern, because such effects may involve not only amplification of reversible changes already known, but also the emergence of heretofore unrecognized and irreversible alterations in cardiopulmonary function. For example, some observers have speculated that there is a loss of cardiac mass during prolonged microgravity exposure. Will lengthy missions render space travelers unfit for return to a one-g environment?

Finally, it should be recognized that the microgravity environment may provide a unique laboratory for investigation of fundamental cardiovascular and pulmonary physiology and development. Basic observations in the microgravity environment may find practical applications to human health on Earth, perhaps in coping with the effects of prolonged bed rest.

Proposed Research

Cardiovascular Conditioning

Several current issues in the area of cardiopulmonary function can be expected to continue to be of great interest 10 years from now, despite investigations now planned for Shuttle missions.

Foremost among current topics is cardiovascular deconditioning. The mechanisms by which bodily fluid is shifted remain to be elucidated, and are undoubtedly not simple. In addition, renal and endocrine changes are likely upon exposure to microgravity, as are neuroregulatory alterations. We need to characterize all of the cardiovascular, renal, endocrine, and neuroregulatory forces at play during spaceflight, reentry, and post-reentry periods.

Postflight orthostatic intolerance is due to more than just loss of fluid. We have evidence that both the autonomic nervous system and hormone secretion are altered. Their effects on the kidneys, blood vessels, and heart have yet to be fully understood and must be studied over varying durations of exposure to weightlessness. Elucidation of the mechanisms of these effects promises to shed light on some clinical, nonspaceflight problems such as high blood pressure and heart failure.

The problem of long-term exposure to microgravity looms large; currently observed space effects may intensify or new ones may appear. All cardiovascular changes now appear to be relatively mild and entirely reversible upon resumption of one-g exposure. Might orthostatic intolerance become irreversible after long-term exposure? How will the time course of cardiovascular readaptation to one-g be affected by lengthier missions? There now appears to be no deleterious effect of spaceflight directly upon the heart; will long-term spaceflight bring irreversible, involutional myocardial degeneration or "hypotrophy?"

It is perhaps surprising that, to date, we have very little understanding of the exact physiologic effects—beneficial or harmful—of various types of exercise on the phenomenon of cardiovascular deconditioning. For example, some evidence suggests that the aerobically trained individual may be more vulnerable to orthostatic intolerance. Protocols for preflight, inflight, and postflight exercise must be designed and tested in a rigorous manner to determine what, if any, types of exercise may be the best countermeasures to deconditioning. Integrated into the problem of understanding the effects of exercise on cardiovascular deconditioning is also understanding the responses of blood gases, electrolytes, glucose, insulin, growth hormone, glucagon, and cortisol.

The microgravity environment will be an ideal place to investigate the neurohumoral control of the cardiovascular system. It is well known that the bodies of upright primates, including human beings, have several "built-in" mechanisms to control blood volume and pressure during postural changes in the presence of gravity. These mechanisms include: (1) changing peripheral vascular tone, mediated by arterial baroreceptors, (2) regulating urine volume and electrolyte output, mediated through low-pressure baroreceptors in the atria, (3) matching of ventilation and perfusion in all lung segments, (4) autoregulating cerebral function, and (5) using check valves in the venous system. These neurohumoral mechanisms are complicated, and involve afferents from the sympathetic system and efferents originating in the hypothalamic and brainstem region and going to the heart, kidney, and peripheral vasculature. The neurotransmitters are numerous, and several hormones, such as the so-called "natriuretic factor," have not been fully characterized.

Work on animals has delineated the importance of these regulatory systems. Dogs whose carotid sinus baroreceptors have been excised have a variation from 40 to 200 in their mean blood pressure. Manual compression of carotid sinuses can cause syncope and bradycardia (low heart rate) in healthy persons. In orbit or in free space, however, the carotid sinus would fire nerve impulses at a more or less constant rate, subject to accelerations generally much less than one-g. The extent to which the sensitivity of the baroreceptors may decrease with prolonged space travel, and their ability to regain the lost sensitivity, is unknown. Given the possibly disastrous consequences of loss of baroreceptor sensitivity, which may have played a role in the syncope and prolonged incapacitation of the Russian astronauts following return from their 211-day mission, it is prudent to investigate the relationship between the duration of microgravity exposure and the magnitude and duration of the loss of baroreceptor sensitivity.

Although we might expect Shuttle missions and earth-based modeling to help, in 1995 we will still need to understand more completely the actions of drugs that affect cardiopulmonary and renal systems in space. This will be essential for adequate health maintenance. Ordered in descending priority, the following classes of agents must be investigated: antiarrhythmics, bronchodilators, antiallergy/antianaphylactic drugs, analgesics (including narcotics), hypnotics/psychotropics, diuretics, and anticoagulants.

Pulmonary Function

Currently, there is little information on pulmonary function during spaceflight. However, it is possible that lengthy alterations in the relative flow distribution of blood and air in different lung regions might permanently affect right heart function.

Dysbarism, the condition that results from exposure to decreased or changing barometric pressure, is a problem of increasing magnitude in the Shuttle period, but threatens to be of even greater importance during in-orbit, manned construction and repair of space stations and vehicles. It is likely that by 1995 we shall still require further, possibly urgent research into dysbarism. This is a problem involving several disciplines, but certainly appropriate to the pulmonary system.

Renal Function

The kidney is central to the above-mentioned physiologic questions. Renal problems may occur in the space environment. As discussed earlier, weightlessness causes a monthly 0.4 percent resorption of bone calcium, which is excreted in the urine. With increased concentration of urinary calcium and some other changes induced by weightlessness (such as urine alkalinity and possible reduction in urine volume), kidney stones may form more easily. In addition to debilitating pain, kidney stones might obstruct the urinary tract and precipitate infection—potentially quite dangerous. Thus, kidney function must be understood better with regard to calcium metabolism as well as its relation to cardiovascular phenomena.

Conclusions and Recommendation

Cardiovascular mechanisms may be interlinked with problems in higher-priority systems: fluid volume shifts may be pivotal in the development of neurovestibular derangements, thus magnifying even further the need to fully investigate the circulatory deconditioning phenomenon and its physiology. In the next 10 years we will learn more about cardiovascular, pulmonary, and renal physiology in the microgravity environment. This will come about through ground-based as well as a limited number of in-flight experiments. However, the new knowledge will not be sufficient to resolve most issues surrounding short-term exposure.

In 1995, it is likely that virtually nothing will be known about the cardiopulmonary and renal effects of long-term spaceflight, particularly the potential for irreversible changes or adaptations. Thus, it is mandatory to incorporate cardiopulmonary-renal research into the general plan of research for 1995 to 2015, with special reference to long-term, incremental exposure.

Methodology and instrumentation for measuring and imaging are expected to be more sophisticated by 1995. We shall continue to make significant strides in several areas including the following: (1) reliability, flexibility, and precision of instruments and methods, (2) miniaturization of instruments, (3) new and more powerful noninvasive imaging, and (4) ability to accumulate and process data on-line, on-board.

The task group suggests several priorities for investigating cardiovascular, pulmonary, and renal systems:

1. We must understand more about exercise: What preflight conditioning or screening programs are really best? What is most effective for preventing and treating deconditioning? What is the optimum postflight reconditioning profile for resuming normal cardiopulmonary health and activity in one-g? What is the role of salt and water loading immediately before reentry, or the use of sodium-retentive drugs or other agents as countermeasures? What, in general, should we plan for the Earth rehabilitation of long-term space travelers?

2. We must validate to a much greater degree our ground-based models of weightlessness, determine their boundaries and validity, and examine their appropriateness with regard to duration of microgravity exposure.

3. We must characterize drug actions and metabolism in microgravity. The aforementioned list of agents must be fully studied to ensure their safe and effective use in space.

4. We must be prepared to conduct long-term experiments as soon as possible to study the possibility of extended or irreversible changes in cardiopulmonary systems. This will necessarily involve observation of animal subjects exposed to weightlessness under completely controlled conditions.

It is important to emphasize that the above investigations will involve, in a crucial sense, the use of animal subjects in space. Such experiments cannot be done without a well-designed, smoothly functioning, and safe animal laboratory; the importance of this concept cannot be overestimated.

INTEGRATED FUNCTIONS

Introduction

Every aspect of human physiology may be affected by spaceflight or extraterrestrial habitation. The preceding four sections of this chapter have addressed systems that have already demonstrated the need for research. In this section, the task group considers nutrition and the immune system. Although every system or process must ultimately be viewed in the context of the entire person, the task group emphasizes the systemic importance of nutrition and immunology.

Nutrition

Background

Prior to the start of the spaceflight program, there was speculation that the decreased effort of movement in weightlessness would result in a diminished caloric requirement compared to that on Earth. Diets were actually planned, however, at caloric levels close to those needed for normal activity on Earth. In practice this procedure has worked reasonably well. In the 1- to 3-month flights of Skylab, modest degrees of body weight loss occurred, associated with body fluid shifts and losses in muscle mass, as astronauts consumed 2400 to 2800 calories per day. There was clearly no lessening of caloric requirements in space.

In the past, many athletes and astronauts have been convinced that high protein intake builds muscle and strength. However, the physiological evidence indicates that protein is increased in muscle only when needed for the muscle hypertrophy required by continuing physical activity; excess calories of any kind are converted to and stored in the body as fat. In addition, numerous previous studies unrelated to space have indicated that increasing the protein intake increases the urinary excretion of calcium. The level of protein in the diets of astronauts, therefore, needs to be reconsidered for its possible relationship to the potential for urinary tract stone formation and possibly to the rate of loss of minerals from the skeleton. Some degree of uncertainty exists as to whether the high phosphate content of meat is partially protective against the effect of high protein intake to increase urinary calcium. At the same time, there must be concern not to accentuate the negative nitrogen balance associated with muscle atrophy in weightlessness by encouraging too low a protein intake. Since negative nitrogen balance in space has occurred at daily protein intakes of 85 to 95 g, the recommended intake should not fall below this level.

Carbohydrates should be of special concern because of their effects on behavior. Abundant evidence supports the view that any dietary carbohydrate that elicits the secretion of insulin can, unless consumed with adequate amounts of protein, increase the synthesis and release of the brain neurotransmitter serotonin. This substance makes people drowsy and interferes with optimal performance. If this relationship is not recognized, menus and the time of consumption of particular items—especially snacks—might not be appropriate to the tasks required, particularly if they are complex and prolonged. It is possible that other food constituents will also be found that affect behavior, mood, and cognition. As carbohydrates are the likely products of future chemical synthetic systems, it is important to determine the type and maximum amount of carbohydrate that should be reasonably contained in a human diet.

The current (1980) Recommended Dietary Allowances (RDAs) of the National Research Council's Food and Nutrition Board list

800 mg as the appropriate amount of calcium to be taken daily with the principal purpose of "protecting" the skeleton. A much higher level is currently under consideration by the committee working on the next edition of the RDAs. In anticipation of approval by the Food and Nutrition Board, 1000 mg is a reasonable base figure for calcium in diet formulation for spaceflight.

Among the countermeasures tested by NASA have been high calcium and high phosphorus intake in both bed rest subjects and Skylab astronauts. The study showed that this procedure maintained calcium intake and excretion level in balance for up to 3 months, following which the gradually rising fecal excretion of calcium caused a negative calcium balance. Hence, there is no basis at this time for recommending a higher intake level than 1000 mg/day.

Bed rest studies of the effects of high phosphorus intake showed some suppression of the tendency of urinary calcium to elevate, but overall phosphorus intake manipulation was ineffective because of gradually increasing fecal calcium excretion. Furthermore, the possible deleterious effect of a phosphorus intake higher than that in an approximate calcium to phosphorus ratio of 1:1.8 must be remembered. Too high an intake of phosphorus will exert some binding effect on calcium in the intestine and tend to inhibit calcium absorption.

The current RDA level for magnesium is 350 mg/day for adult males. While studies of this element in relation to bone are far less numerous than studies of calcium, research to date indicates that deleterious effects apparently do not occur except possibly with low intake—as in an artificial diet—over a very long time.

Since no studies have yet been made on the effects of spaceflight on the metabolism of any of the trace elements, no comment can be made other than that care should be taken that space diets contain trace elements in the amounts recommended in the RDAs.

The important vitamin in long spaceflights is vitamin D, the "sunshine vitamin." Enclosure in a space vehicle will prevent the normal conversion in the skin of the vitamin D precursor to vitamin D. This is normally accomplished by exposure of the face and arms to as little as 20 to 30 minutes of sunlight a day. Since vitamin D is essential for facilitating calcium absorption from the intestine, as well as other calcium-related effects in kidney and bone, this vitamin will need to be supplied to space travelers.

However, amounts should not exceed about 800 to 1000 I.U. per day.

Other vitamins are not so critical since it is expected that adequate amounts will be taken in the diet, provided it is "balanced" and the vitamins are not degraded by the methods of food preservation in use. It has become customary, however, to provide astronauts with daily vitamin supplements.

The absence of natural light in spacecraft may have significant effects other than that on vitamin D synthesis. For most of man's evolutionary history he spent his days out-of-doors, exposed to 1000 to 8000 foot-candles of light provided by the Sun's rays (filtered through the ozone layer), including a small but important amount of mid- and near-ultraviolet light, and approximately equal portions of the various colors of visible light. Indoor lighting in most offices and, so far, in all spacecraft is of a much lower intensity (usually 60 to 100 foot-candles), and, if emitted by fluorescent "daylight" or "cool-white" bulbs, is deficient in ultraviolet light (and the blues and reds) and excessive in the light colors (yellow-green) that are best perceived as brightness by the retina. If light's only effect on humans was to generate subjective brightness, then this artificial light spectrum might be adequate. It has become abundantly clear, however, that light has numerous additional physiological and behavioral effects, and that the "action spectra" of these effects (i.e., the extent to which they are activated by different wavelengths) differ from the brightness spectrum. For example, light exerts direct effects on chemicals near the surface of the body, photoactivating vitamin D precursors (as noted above) and destroying circulating bilirubin and probably other photoabsorbent compounds. It also exerts indirect effects via the eye and brain on neuroendocrine functions, circadian rhythms, secretion frcm the pineal organ, and, most clearly, on mood. Many people exhibit major swings in mood seasonally, veering toward depression in the fall and winter when the hours of daylight grow short. Some cross into acknowledged depression, now known as the "seasonal affective disorder syndrome," a disease that may be related to excessive secretion of the pineal hormone, melatonin, and which also may be treatable with several hours per day of supplemental light. While not yet proved, it seems highly likely that prolonged exposure to inadequate lighting—that is, the wrong spectrum, or too low an intensity, or too few hours per day of light—may adversely affect mood and performance. Available information, though not fully satisfactory, suggests that the lighting environment now provided in spacecraft may indeed be inadequate. The adequacy of lighting should be part of the planning for long-term spaceflights, and all physiological, biochemical, and behavioral effects of light should be studied extensively.

In the early days of planning for manned spaceflight, many thought that diets should be low-residue in character so that bowel movements would be small and infrequent. It was observed especially in longer flights that bowel function in microgravity is essentially normal. Hence diets should be normal in residue, and adequate bulk should be available to afford relatively easy passage of stools once or twice a day.

Proposed Research

In spaceflights extending from many months to years, the acceptability of various currently available packaged, canned, freeze-dried, or heat-stable food items should be evaluated. As the capacity to carry and store frozen food items is likely to be limited in extremely long flights, space food technology research should be revived in planning for the Space Station era. To date, nutrition investigations (unrelated to space) suggest that individuals do not crave continual variety in foods but rather tend to select foods in the same small range or limited number over months, stretching out to a lifetime. Reduction in the total list of available food items should also simplify both the strategy of storage of multiple food packages in a long-flying spacecraft and the ability of travelers to pull out desired items with a minimum of difficulty and time. The Space Station will need to provide the function of testing the long-term durability and acceptability of both currently available and newly formulated items.

There will also be a need to check out current calculations that there will be adequate and satisfactory food storage on long-flight vehicles for 3 years' flight duration. These studies should provide guidelines for the CELSS program. In turn, much of the research in nutrition should be guided by realistic projections of the characteristics and quantities of food to be produced by CELSS.

Immune System

Background

Although the reports to date are conflicting, some indicate that a microgravity environment may compromise the immune system's function. Studies on cell-mediated immunity of 21 crew members on Apollo missions showed that 7 did not reveal any consistent changes. In contrast, Cogoli et al. reported that cultures of human lymphocytes subjected to microgravity responded to concanavalin A, a lymphocyte stimulating agent, 97 percent less than ground-based controls. This is a standard test used to evaluate the competence of peripheral blood lymphocytes to multiply when stimulated with this agent. Studies on the astronauts of the first four STS flights revealed that the lymphocyte responses to photohemagglutinin, another lymphocyte stimulating agent, were reduced from 18 to 61 percent of normal following spaceflight. It has been suggested that the above changes were due to stress-related effects, but this has not been established and should be studied further.

In an unmanned Russian spaceflight, it was reported that rats flown for 22 days had marked reduction in the weights of lymph nodes and spleens compared to controls on Earth, due to a marked decrease of lymphocytes in these organs. The effects were found to be reversible since the organs returned to normal by 27 days postflight. In another study, Mandel and Balish studied rats subjected to a 20-day flight aboard the unmanned U.S.S.R.-Cosmos 7820. They immunized groups of rats with formalin-killed *Listeria monocytogenes* 5 days before flight, and compared animals exposed to space conditions with one-g controls. They concluded that no deterioration of the acquisition cell-mediated immunity to *L. monocytogenes* could be detected in flown rats.

In one study, mouse or human lymphocytes subjected to 2-g and 4-g, respectively, exhibited enhanced responses to concanavalin A compared to parallel cultures at one-g. In other investigations, hypergravity had little or no effect on the humoral response. On the other hand, graft survival time was increased in rats subjected to 2.5-g for 4 days preoperatively and 3-g postoperatively. These findings indicate that cell-mediated immunity against tissue grafts may be compromised in hypogravity. These studies also emphasize the need for one-g inflight controls of immune system investigations.

In view of the present uncertain status of the effect of hypogravity and hypergravity on immune function, the task group recommends that the basic components and function of the immunologic system be studied systematically at enhanced and diminished gravity. These studies assume added importance because, as discussed earlier, the concentrations of microorganisms in space vehicles may be significantly higher than normal.

Proposed Research

Specific questions to be answered by studies in microgravity include the following:

1. What are the effects of microgravity on the lymphoid organs—nodes, spleen, thymus, and bone marrow?
2. Is the response to antigen priming altered?
3. Is the response to antigen-induced secondary response altered?
4. What are the induced lymphocytic subsets?
5. What is the receptor density on macrophages including C3b, Fc, and Ia under conditions that activate macrophages?
6. Is the formation of the allergic granuloma and delayed hypersensitivity response using BCG strain of *Mycobacterium tuberculosis* altered?
7. What are the serum Ig and complement concentrations?
8. What are the interferon and interleukin 2 concentrations?

Although there is need to answer these questions for the human system, some experiments must be done on small mammals.

The conditions associated with space travel, space stations, and planetary colonies raise many new and important problems concerned with host-parasite interactions involving man and animals. Rotation of crew members on the Space Station will introduce different strains of fungi, bacteria, and viruses that could contribute to the emergence of "new" strains of opportunistic pathogens through mutation and genetic exchange. Investigations should also consider the dynamics of aerosol generation and microbial survival under weightless conditions. As discussed in Chapter 8, special equipment for these studies includes: an image transmission microscope, a laser cytofluorograph, an immunoelectrophoresis device, and a small 20,000-g centrifuge.

Other Systems

At this time it is not possible to certify any physiological system to be unaffected by several years at microgravity or to preclude any as a fruitful area of research. Preliminary results indicate reduced hematocrit in some astronauts, but this may be a physiological readjustment that is appropriate. Whether hematopoesis or maturation of lymphocytes is compromised is yet to be established. The multiple stresses of spaceflight may lead to hormonal imbalances. Corticosteroid release may lead to immunosuppression. Oogenesis and spermatogenesis may be compromised. In any case, additional research is required to confirm or reject the presence of problems.

At present, we cannot assume that as spaceflight increases from months to years unanticipated malfunctions will not appear. We must continue to establish a reliable data base so that we can recognize and research these new phenomena before proceeding to longer flights. To accomplish this, we must continue to employ the approach of incremental exposure of humans to microgravity with careful surveillance during and after exposure.

RADIATION EFFECTS

Introduction

The importance of the radiation factor has been underscored by several committees of the National Research Council and other NASA-sponsored committees. The evolving interest in prolonged manned space travel beyond the Earth's protective magnetic field brings to the forefront uncertainties in the physical behavior and biological effects of the so-called "free space" radiation environment. It is generally agreed that these uncertainties must be resolved before we embark on the construction of a lunar base, the manned habitation of space platforms, manned flight to Mars, or lunar or martian habitation.

Background

Space Radiation Environment

There are basically three sources of naturally occurring space radiation that can be hazardous to manned spaceflight: the geomagnetically trapped proton and electron environment (Van Allen belts), galactic cosmic radiation (GCR), and solar particulate radiation.

The Van Allen belts consist of high-energy protons (approximately 1 keV to several hundred MeV) and electrons (approximately 1 keV to several MeV) trapped in the geomagnetic field. The proton belt extends to an altitude of approximately 20,000 km, with peak intensities occurring at approximately 5,000 km. The electron belts extend to an altitude of 30,000 km, with peaks at about 3,000 and 15,000 km. Models of the trapped proton and electron environments have been developed from satellite measurements.

Galactic cosmic radiation (GCR) consists of extremely energetic (up to 10^{13} MeV) ionized nuclei ranging from hydrogen to uranium and originating outside the solar system (so-called HZE particles). Models of the GCR environment have been generated from geostationary satellite and high-altitude balloon measurements. Current knowledge of the GCR hazard, however, is inadequate because of the poor understanding of the effects of HZE particles on biological tissue.

Solar particulate radiation (solar particle events) consists of high-energy particles (predominately protons) ejected from the Sun, usually during solar flares. Solar activity has an 11-year cycle, during which a tenfold variation in the frequency of particle events has been observed. No reliable physical model can predict the timing or magnitude of solar particle event (SPE) occurrence with acceptable accuracy. This feature makes SPEs a significant hazard in long-duration space travel. Additionally, solar flare activity can substantially increase the fluence of HZE particles, at least up to energies of a few hundred MeV per nucleon. The high-altitude IMP 8 satellite, for example, observed 28 solar flares during solar cycle 21 in which the flux of heavy ions was substantial. At 100 MeV/nucleon the fluxes of carbon and oxygen ions were approximately 10 times the ambiance of the GCR flux. Occasionally flares are observed to be iron rich, and a flare in 1977 produced a fluence of 200 MeV/nucleon iron ions 10 to 20 times the GCR fluence. As will be mentioned presently, the absorbed dose in biological tissue in these events is not negligible, and the effects on spacecraft electronics could be significant.

Radiation exposure in low earth orbit (LEO), where Shuttle orbits lie and the Space Station orbit will lie, is primarily from the proton and electron belts and GCR. Trapped-radiation exposure increases with altitude and varies with orbital inclination. GCR exposure varies with orbital inclination: from approximately 5 mrad/day at 28° to approximately 20 mrad/day for polar orbit during solar minimum, and approximately 3 mrad/day at 28° to approximately 15 mrad/day for polar orbit during solar maximum. The geomagnetic field provides some degree of protection from solar particle events, depending on the orbital inclination; flux is almost totally eliminated for a 28° orbit and reduced to about 30 percent of the free space flux for polar orbit.

Exposure at geosynchronous (GEO) altitude will be primarily from bremsstrahlung (x rays) created by the trapped electrons as they interact with spacecraft shielding. The electron environment at GEO has a diurnal fluctuation, and intensities can increase by several orders of magnitude with magnetic storm activity. GEO is susceptible to the full exposures from GCR and solar particle events, as are lunar and interplanetary missions.

Spacecraft Radiation Environment

Incoming radiation from space is modified as it passes through the body of a spacecraft and any additional shielding that may be present. The biological effects of radiation must be determined, therefore, by starting with this modified spectrum. The physical principles by which radiation interacts with matter are well known, but the way to combine these principles to form a good model of the resulting secondary spectrum is not. Models for the magnitude of the trapped radiation in certain energy ranges have uncertainties ranging from a factor of 2 for the inner belt to a factor of 10 or more for the outer belt. In addition, the trapped radiation models were developed in the early 1970s and are in need of refinement. Nuclear fragmentation cross sections for heavy ion constituents of the GCR are, in some cases, completely unknown. A substantial amount of data obtained from various forms of dosimetry onboard Apollo, Skylab, and STS missions has provided measurements of radiation exposures, but these data cannot be extrapolated to free space. Nevertheless, with available models and limited spacecraft data, the daily exposure for various mission configurations has been estimated. For the Space Station, the dose to the blood-forming organs (BFO) has been estimated to be approximately 100 mrad/day, of which approximately 90 percent will be from trapped protons. During as SPE, the absorbed dose would be mainly from high-energy protons. The SPE of August 1972 would have produced approximately 150 rad to the BFO for a mission in free space assuming reasonable shielding, such as that provided by the STS orbiter. (The exposure would have been uniformly lethal without shielding.) High charge and energy ions (HZE) from the GCR contribute about 30 mrad/day in free space, independent of the amount of shielding. Indeed, the inability to shield effectively against the GCR in free space will be a persistent problem for long-duration missions to planets, for platform habitation, and at a lunar or martian colony. As stated previously, solar flare activity can result in heavy ion fluences at a few hundred MeV per nucleon that are up to 10 times higher than the background GCR flux. This fluence may result in a 24-h exposure of up to 300 mR of high linear energy transfer (LET) radiation. Depending on the relative biological effect of these ions, the dose could be significant.

Biological Effects

The biological effects of ionizing radiation have been extensively studied for almost a century. The data come from studies of controlled irradiation of cell cultures, small and large animals, and nonhuman primates, as well as from retrospective studies of humans exposed to nuclear weapons blasts, radiation used for medical treatment, and nuclear occupational hazards. Most of the information has been obtained with so-called low-LET radiation such as x, gamma, and electron radiation. Low-LET radiation is sparsely ionizing—it is characterized by separated clusters of ionizations along the path of the primary photon or electron. In contrast, high-LET radiation, such as stopping protons, secondary stopping protons from neutrons, alpha particles, and energetic heavy multicharged particles, is densely ionizing.

It has been known for decades that a given amount of energy deposited by high-LET radiation could be several times more damaging than the same amount of energy deposited by low-LET radiation. Because of the higher relative biological effectiveness (RBE) of high-LET radiation, a quality factor (Q) is applied to occupational doses (in physical units) to obtain a weighted unit for assessment of radiological health risk (dose equivalent). For example, the Q for neutrons from a nuclear reactor would be about 10. The International Commission on Radiation Protection (ICRP) has a long-pending recommendation that calls for a Q ranging from 1 to 20 depending on the LET of the particle. In recent years, however, evidence is mounting that under certain practical circumstances, RBEs can be 40 to 100. These circumstances include neutron radiation, low doses at low dose rates; and certain biological endpoints, such as effects related to cancer induction (chromosomal abnormalities and rearrangements). This recent information has led to a ferment in the radiological health community, and Q values higher than 20 are currently being proposed. More generally, the assumed linear relationship between absorbed dose and observed biological effect has come into question for HZE particles or high-LET particles in general. Since the manner in which energy is deposited in tissue by HZE particles is so different from that of low-LET particles, this linearity may not apply to HZE particles. Individual physical circumstances in the way in which these particles interact with various biological tissues must be considered, rather than including all such interactions in a single ratio called "relative biological effect." This opinion has been expressed both by the National Council on Radiation Protection and Measurements and by prominent high-energy particle research groups. Of current interest has been the "microlesion" concept. This theoretical model of the interaction of heavy particles with biological tissue has raised the question of a whole new spectrum of biological damage, including damage to nondividing cells, particularly the central nervous systems. It appears that the microlesion concept is worthy of further investigation, as there may be significant consequences in long-duration spaceflight ($\geq$ 3 years) if an accidental underestimation of the effect of HZE particles is made.

The assessment of the radiological health risks for various future missions (Space Station, polar orbit, GEO sorties, lunar base, and Mars missions) and thus the operational limits for such missions are dependent on Q, which in turn will be greatly dependent on the evaluation of RBE using relevant biological criteria (life shortening, tumor induction, chromosome abnormalities, mutation, teratogenesis, and so on). The data base using space-type radiation for such assessments is disturbingly small.

Estimates of carcinogenic risk have been made by the National Institutes of Health Ad Hoc Working Group to Develop Radioepidemiological Tables. These risk values have been expanded upon for use in the report of Scientific Committee 75 of the National Council on Radiation Protection and Measurements (NCRP) entitled *Guidance on Radiation Received in Space Activities*. Table 6.2 (extracted from the NCRP report) shows the best currently available estimates for the effects of 1 sievert (Sv)* of radiation spread over 10 years. Although the values are presented to three significant figures, it should be emphasized that there are large uncertainties. As further statistics on the Hiroshima-Nagasaki survivors become available, these values are expected to change somewhat. NASA must continue to monitor developments in such radioepidemiological efforts. Nonetheless, these radiogenic cancer risk estimates given in Table 6.2 have served in part as the basis for the new set of astronaut radiation exposure limits being recommended to NASA by NCRP. These new limits are shown in Table 6.3. The limit of greatest importance for future space activities is the career dose-equivalent for the BFO. For example, the BFO career dose-equivalent limit for a 30-year-old male or a 38-year-old female is 2 Sv. In practice, it will be unlikely that any astronaut will receive 1 Sv over a career (barring exposure to an unanticipated dose from a large and unexpected solar particle event). However, the 1 Sv value will be approached during a 2- to 3-year Mars mission, given currently used quality factors. If the Q for GCR heavy ions increased, the 1-Sv level may very likely be exceeded. The career limits are not expected to be raised, because the risk estimates are based on radioepidemiological data from humans exposed to low-LET radiation for which the Q is unity. In other words, the report of Scientific Committee 75 of the NCRP entitled *Guidance on Radiation Received in Space Activities* did not give any consideration to the effect of HZE particles. As we move into the future toward lunar colonization and long-term

---

*A sievert (Sv) is the SI unit for dose-equivalent, which is the physical dose in Gray (Gy: 1 joule/kilogram) multiplied by a quality factor (Q) to account for the increased biological effectiveness of some radiations. The Sv is equivalent to 100 rem; the Gy is equivalent to 100 rad.

TABLE 6.2 Cancer Morbidity and Mortality by Age Group and Sex, With and Without Radiation

Sex	Age	Lifetime Incidence (%) Unirradiated	Additional Incidence (%) from 1 Sv[a]	Lifetime Mortality (%) Unirradiated	Additional Mortality (%) from 1 Sv[a]
Male	25	34.9	3.10	18.5	1.99
	35	35.2	1.84	18.7	1.20
	45	35.5	1.38	18.9	0.92
	55	35.4	1.12	18.7	0.75
Female	25	35.6	6.24	15.7	2.93
	35	35.2	3.50	15.5	1.70
	45	33.9	2.22	15.1	1.19
	55	30.8	1.73	13.9	0.99

[a] 1 Sv at 0.1 Sv/year for 10 years starting at indicated age (1 Sv = 100 rem).

TABLE 6.3 Astronaut Ionizing Radiation Exposure Limits[a]

	Depth (5 cm), cSv[b]	Eye (0.3 cm), cSv[b]	Skin (0.01 cm), cSv[b]
30 Days	25	100	150
Annual	50	200	300
Career	100-400[c]	400	600

[a] These dose-equivalent limits are being recommended to NASA by NCRP Scientific Committee 75 on Guidance on Radiation Received in Space Activities.
[b] cSv = centisievert, which is based on the SI unit of dose equivalent, the sievert (Sv); 1 Sv = 100 rem; 1 cSv = 1 rem.
[c] The career depth dose equivalent is based upon a maximum 3 percent lifetime risk of cancer mortality. The total dose equivalent yielding this risk depends on sex and on age at the start of exposure. The career dose equivalent limit is approximately equal to 200 + 7.5 (age 30) cSv, for males, up to 400 cSv maximum; and 200 - 7.5 (age 38) cSv, for females, up to 400 cSv maximum.

interplantary travel, which might involve entire lifetimes or perhaps even generations of lifetimes in space, entirely new standards will have to be devised to account for this dominating source of radiation.

The possibilities of modifying the biological damage by radiation deserve further attention. Recent evidence obtained by the cancer research community indicates that the multiphase process of carcinogenesis can be interrupted at various stages. For example, at the DNA damage or initiation phase, free radical scavengers, such as vitamin E and possibly vitamins A and C, can protect. Some data indicate that the promotion phase, in which a radiation-damaged cell changes to a potentially cancerous cell cluster and then goes on to the progression phase yielding a tumor, can be interrupted by agents such as dimethylsulfoxide (DMSO) or protease inhibitors. Implementation of the results of studies directed toward early detection of cancer could help improve the prognosis for crew members unfortunate enough to contract cancer.

Before a commitment to a lunar or martian colony is made, mutagenesis and teratogenesis by high-LET radiation must be extensively studied. Mutation and developmental abnormalities are, like cancer induction, stochastic effects: the severity of the effect is independent of dose, but the probability of occurrence increases with dose. The mutation risk to future generations from expected space radiation doses is apparently fairly low, but the available information is woefully inadequate for assessing the teratogenic risks to fetuses.

Although most space radiation doses will be low and received at low dose rates, a very large solar particle event can expose astronauts in polar orbit, in GEO, or in free space to high-dose and high-dose-rate radiation, which can produce clinically significant effects. These effects are nonstochastic: the severity of the effect increases with dose above some effective threshold. The acute radiation syndrome (ARS) at sublethal doses is characterized early on by transient anorexia, nausea, vomiting, and diarrhea. Later, the survivors may suffer temporary or permanent sterility and cataracts, as well as cancer. Lethal doses lead to bone marrow suppression and immune system compromise, which leads to hematopoietic death in 30 to 60 days. These high doses lead to severe gastrointestinal disturbances in 1 day to 1 week. Extreme doses can produce central nervous system derangement in a matter of hours.

The ARS has been studied extensively over the past four decades and continues to be studied in animal models by, for example, the Armed Forces Radiobiology Research Institute (AFRRI). The human clinical experience, however, has been extremely limited. It is important to stress that the treatment of ARS has been largely supportive. This limitation implies the need for a "critical case" of an astronaut exposed to potentially lethal doses of radiation in a solar flare or onboard nuclear containment accident. Onboard provisions for such management must be based on the philosophical decision to treat severe illness in space travel—a decision that has not yet materialized in the U.S. space program.

In the prevention of high-dose acute radiation exposure, special shielding is the most commonly considered modality. In situations where it is not possible to deorbit or lower the altitude to a protected region of space, a storm shelter with adequate shielding must be provided. For example, a water-filled collapsible cocoon has been proposed. For added protection in case a very large solar event occurs, partial body shielding of a small amount of bone marrow stem cells can be very effective in raising the lethal threshold: for example, in one study monkeys that had 1 percent of their stem cells protected survived a dose that killed all unshielded animals. In the future, ex vivo cell storage techniques may allow a bank of shielded autologous bone marrow to accompany astronauts on a long-duration mission.

Proposed Research

As we embark on an era of the U.S. space program in which lunar colonies and manned Mars missions are being seriously discussed, we are largely unaware of the possibly serious consequences of both long-term exposure to the free space GCR and short-term exposure to an SPE.

The task group recommends the following major research, beginning now and continuing in the years 1995 to 2015:

1. The biological dosimetry of high-LET radiation should be developed and formalized to allow for studies of comparative risks.

2. Shielding models applicable to the GCR spectrum should be further developed. Emphasis should be given to realistic shielding materials and thickness suitable for manned spacecraft.

3. A free space satellite carrying GCR detectors should be sent with the intent of fully characterizing the spectrum behind a realistic shield in free space.

4. Terrestrial studies of the biological effects of low-level, high-LET irradiation on cell cultures and animals (using particle accelerators) should be expanded, with particular attention paid to the space radiation problem.

5. Characterization and prediction of solar particle events should continue, with an awareness by solar physicists and astronomers of the biological relevance to the future of the U.S. space program. In particular, the directionality of particles in low-altitude polar orbits needs characterization in order to optimize the use of available spacecraft mass for shielding.

6. Methods for the critical care of acute radiation injury should be addressed, with a focus on manned space mission application. In the spacecraft setting, additional innovations are possible, such as carrying onboard shielded, frozen, autologous bone marrow or partial body shielding of blood stem-cell-rich body regions.

7. Studies of drug interventions should proceed.

BEHAVIOR AND PERFORMANCE

Introduction

Nearly all essential human functions during long-term space missions will depend critically upon individual and group behavior. To date, selection, training, and organizational functions have been the focus of human behavioral initiatives. A program of fundamental scientific research treating behavior interactions is as essential to ensure the success of long-duration human operations in space as are the physical science and engineering investments that make such initiatives possible.

Before the long-term occupancy of space environments can be safely and productively assured, we must address the historical problems of individual behavioral adjustment, interpersonal conflict, and group performance effectiveness. That such obstacles are typically exacerbated in isolated and confined microsocieties has been repeatedly documented in operational settings such as remote stations in the Antarctic, undersea habitats, and most pertinently, in spacecraft. To some extent, these earth-based experiences provide relevant data to questions regarding human behavior and performance in space. However, the fact that observations are made and observed by people who actually share the experience limits the reliability of data. The conditions that will exist in long-duration space missions increase the potential for adverse effects already reported during relatively short-term missions (e.g., irritability, depression, sleep disturbances, and poor performance of both group and individual).

It is essential that we develop the scientific foundation for the provision of adequate individual, social, and organizational environments. Long-term space travel will require establishing and maintaining effective, stable interactions between individuals in small groups that are under microgravity conditions and are isolated and confined for prolonged periods. However, the research data base pertaining to these conditions is extremely limited. Behavioral and social problems are regarded among those who have been involved in extended space missions to date as formidable obstacles to long-term voyages. Even more important from a life science perspective, it seems likely that entirely new principles of human interaction and group dynamics will emerge as a result of such research to ensure effective human behavior in space environments.

Background

The current status of knowledge in the field of human behavior has provided for substantial progress toward the goal of extended space occupancy. Careful attention to selection, training, and organizational functions has permitted small groups of individuals to live and work effectively in space for continuous periods of several months. But there are enormous gaps in our understanding of how the multiple, complex behavioral factors operate independently to influence the behavior of individuals and groups. While these deficiencies are generally recognized, it seems unlikely that substantial progress in this important field will be under way, much less achieved, by 1995. Deficiencies in understanding the dynamics of the performance and psychosocial health of groups, and especially the integrative factors that ensure the effective function of the system, will be particularly evident.

Proposed Research

Overview

A description of the research required in the realm of human behavior encompasses a range of experimental questions related to the establishment, maintenance, and enhancement of human productivity under progressively autonomous conditions.

Many construction and observation tasks will be done under extremes of pressure, temperature, and radiation hostile to the human body. Human judgment and ingenuity are often indispensible to the completion of some of these jobs. This will demand optimal interactions between humans, machines, and computers. This should not be seen as a choice between man and robot, but as a challenge to integrate judiciously their respective capabilities. Specifically, the primary research objectives of the human behavior studies the task group recommends are:

1. To analyze the environmental factors and associated task requirements that determine completion of mission objectives and enhancement of creative accomplishment.
2. To understand the individual factors and the interactive physiological/behavior processes.
3. To characterize the group factors that influence maintenance of performance effectiveness and enhancement of behavioral competence in space environments.
4. To determine the integrative factors that ensure the most effective systems management of essential environmental, individual, and group requirements in space.
5. To evaluate the design of robots with a goal of optimizing human performance in unique conditions.

Environmental Factors

If long-term space missions are to become a reality, automation, robotics, artificial intelligence, and other advanced computer technologies will increasingly dominate both the physical and the behavior features of space environments. This will necessitate empirical determinations of optimal "mixes" between human control and control by computerized management systems over the entire range of life support, work task, and general performance functions. The development of an experimental base to ensure effective workplace designs under confined, microgravity conditions and to provide for creative integration of living accommodations should be the first step in this program of research.

Individual Factors

Perhaps the most important research priority in considering the impact of individual factors upon personal adjustment to occupancy of space environments is the screening and selection of prospective participants. For the most part, screening efforts have emphasized identification and elimination of potentially disruptive individuals and selection of those individuals evidencing the highest proficiency. The dilemma of choice between achievement and interpersonal harmony is likely to become particularly prominent in the isolated and confined microsociety of the space environment.

Of primary concern must be the development of more valid and reliable methods for observing and recording the effect of space stresses upon complex performance processes. The scientific literature has documented the effectiveness of trained participant observers. A closely related area of research focuses upon the development and refinement of behavioral procedures for physiological self-regulation (e.g., biofeedback techniques).

Important questions surround the general health performance effects of shifting circadian periodicities and extended exposure to drastically modified sensory environments. A host of specific research questions have been raised documenting the intimate relationship between biological rhythms and behavioral interactions. A rapidly emerging behavioral pharmacology data base suggests that highly specific drug performance interactions hold promise for the facilitation and stabilization of behavior under a broad range of environmental conditions. A determination of the extent to which pharmacological interventions can be of benefit in the management of such space-related adaptational problems requires experimental analysis.

Sensorimotor and perceptual functions of long-term isolation and confinement also require concentrated investigation. Training and testing procedures with advanced subhuman primates have been documented. They suggest that one of the more fruitful applications of an animal experimental flight program would be obtaining reliable behavioral measures of sensory thresholds and motor function over extended periods of up to several years. This is one of the most difficult research questions associated with the long-term effects of continuous exposure to space environments.

Perhaps the area of greatest long-range scientific promise is that of an experimentally derived, functional account of individual behavioral variability over time. Without such knowledge, a natural science of behavior cannot exist. Without a natural science of behavior, the social sciences, to which we turn for guidance in the establishment and maintenance of stable behavioral ecosystems in space, will continue to be of limited use.

Also of prime importance are the motivational issues raised by the prospect of long-term space occupancy. Motivation plays a critical role in maintaining individual performances and amicable social interactions over extended intervals of isolation and confinement. It is essential that behavioral research explore the extent to which such motivational processes can be identified in the restricted environment of space travel. This investigation will require that the fundamental structure and function of the human biological and behavioral repertoire be monitored, measured, and analyzed experimentally under simulated conditions involving small groups in isolation and confinement.

Group Factors

Despite an extensive literature on small group structure and function, it has become clear that our understanding of in these important areas is not adequate for spaceflight program planning. There is increasing evidence that groups are, in fact, small social systems shaped by multiple determinants no one of which, considered in isolation, can necessarily account for the variations in behavioral interactions or performance effectiveness.

Much remains to be learned about the partitioning of authority and autonomy among (1) group managers at the base of operations, (2) group leaders internal to the operation, and (3) individual group members. Existing research findings are unambiguous in showing that a clear, engaging set of objectives that "stretch" a group is a powerful means for orienting members toward achieving overall organizational goals. Little is known, however, about the relative effectiveness of alternative strategies for "charging" a group, or about the means of reinforcing overall direction in the course of actual performances, particularly for groups that are physically distant from the central base of operations for extended periods of time. Moreover, additional research is needed on the appropriate exercise of authority in managing the inevitable problems and disputes that occur in real time and that threaten the overall integrity of the group.

While a reasonable knowledge base exists in the areas of selection, placement, and training of individuals for solo task assignments, the data on group composition and task design are rudimentary. For collective operations, the right people, well trained and properly configured (that is, with the right mix of skills, personal characteristics, task requirements, and work setting), are essential.

An important yet poorly defined area bearing upon an understanding of group performance effectiveness concerns competent leadership. A promising new approach to leadership research involves focusing on the identification of those functions that leaders perform in enhancing effectiveness and efficiency.

Integrative Factors

The goal of providing a scientific foundation for human behavioral transactions in space environments requires integration of the separately considered environmental, individual, and group factors within an organizational and systems management context. It is becoming clear that these problems are both interrelated and interactive. We must develop a research analysis model specifying the major integrative features of organizational management systems that foster high performance effectiveness.

One of the more important integrative problems is the relationship between formal organizational structures and the emergent social structures that serve as the milieu for daily life in an isolated and confined microsociety. The accommodation of leisure time activities and the need for individual privacy are but two of the more salient integrative issues deserving research in this regard.

Technology and Scientific Resource Requirements

The task group recommends the development of methods and instruments for the structural and functional analysis of vocal utterances of crew members. This may be especially valuable in evaluating the moods of astronauts or as an early warning system of performance degradation and interpersonal conflict.

The task group also proposes the creation of a long-term residential laboratory providing for the study of behavior under continuous environmental control. The laboratory must simulate the anticipated requirements of future long-duration space operations by incorporating the following features or activities:

1. Environmental design: the facility must contain at least some approximation of the physical features of anticipated space vehicles, space stations, or extraterrestrial colonies.
2. Programmable control of environmental resources, such as temperature, light, food, and recreational facilities.
3. Inventory of essential behavioral activities, such as sleep and waking, eating and drinking, personal hygiene, work, recreation, and social interaction.
4. Biochemical, physiological, and behavioral monitoring of certain biological functions, such as changes in endocrine, autonomic, and skeletal processes.

The task group recommends that analogue research settings, where it is possible to manipulate environmental, group, and organizational factors that bear on spaceflight success, be vigorously supported as an important source of data on human behavior science. Perhaps the closest operational analogue of space occupancy is the undersea habitat, where aquanaut divers live and work on the ocean floor with a degree of isolation similar to that in space. Under these circumstances, and in Antarctic stations and submarine operations as well, observational measurements have focused upon critical individual and group factors that influence performance effectiveness and interpersonal relations. While these analogue studies lack the control of laboratory experiments, the behavioral interactions involved have much in common with space habitation. Thus, these studies can be one of several elements in the vigorous ground-based program in human behavioral research necessary to support any manned mission of long duration.

HEALTH MAINTENANCE

Introduction

As humans establish a permanent presence in space, whether it be on a space platform or in lunar or martian colonies, it is imperative that health care be provided to workers, scientists, and astronauts. The required facilities, procedures, and expertise will demand development of new technologies that pay special heed to the constraints and unique stresses of space, and to the new findings anticipated in the preceding chapters.

Another essential aspect of a health maintenance facility is its interrelation to other life sciences activities. Experience over the past century in the development of modern medicine has shown a strong correlation between optimal medical care and scientific investigation. This concept should be extended into extraterrestrial medicine. To do so should positively affect not only the quality of care but also the quality of life sciences research.

Types of Care

The types of care that a health maintenance facility must provide on a minimal basis fall into four categories: prevention, treatment of disease, treatment of injury, and rehabilitation.

One of the most important components of a health maintenance program is prevention, that is, the maintenance of physical and mental health. Considerations include: physiological status monitoring, nutrition and stress management, safe waste management, hygiene, medical record keeping, environmental monitoring, exercise machinery and facilities, assurance of a suitable sleep environment, recreation and entertainment, social support aids, and communication with family and friends. We must establish physiological norms for both a space station and long-term missions.

For missions of a few months' duration, the trade-off between the capability for emergency transportation back to Earth and the capability for emergency treatment in space must be studied. If emergency rescue is found to be impossible or impractical, then emergency care capability must be improved. It may be necessary to provide a physician for inflight care. Such a person could also be a trained astronaut capable of performing other duties including life sciences research. On long-duration missions, e.g., at a lunar base or on a Mars mission, the need for such personnel will increase greatly. It is imperative that these physicians have access to consultation with other medical specialists on Earth. Cross-trained individuals could provide surgical assistance, anesthesia support, and diagnostic capability, such as in the laboratory or imaging areas.

The concept of a "safe haven" is imperative in planning any

health maintenance facility. Such safe havens could provide temporary protection against fire, environmental toxins, decompression, and radiation. In addition to prevention and treatment, rehabilitation should be considered so as to enhance optimal crew productivity and return to operational capability.

Although space health care should ideally equal terrestrial care in quality, the cost of this care and the expected evolution of what is currently an untested endeavor must be considered. Previous studies at the Johnson Space Center have outlined a four-tiered system of health care that would be adaptable to a space station or even a lunar or Mars colony. The first system in health care can be roughly categorized as a facility that would provide simple first aid, with one or all members of the crew trained in basic care. Equipment would be minimal and would not include integration of life sciences research with the health maintenance facility.

A second-tier health maintenance facility would be a dedicated area for first aid and exercise. In addition, equipment for treatment of hypobarism might be considered. The objectives would be to stabilize the injured patient until rescue could occur, treat minor injury, and even carry out some minimal invasive diagnostic studies and simple diagnostic testing. Such a facility would require extended training of a crew member. Symptoms and clinical signs could be described to physicians on the Earth, who would direct treatment giving instruction to the paraprofessional in the space station or space colony. This seems to be a minimum requirement for a space station.

Injury is the most likely debilitating or potentially life-threatening process, if personnel are young, healthy individuals. There are, however, certain medical and surgical emergencies that affect even young people, such as appendicitis, perforated ulcer, renal stones, and subarachnoid hemorrhage.

In the Polaris submarine missions there have been approximately 21 cases of appendicitis—17 of which were successfully treated with antibiotics and 4 of which resulted in death. Surgery is the only alternative when antibiotics fail and is the primary treatment on Earth in a nonremote setting. If the health maintenance facility is incapable of providing surgical care, the workers and scientists, as well as the public must be aware of such a conscious decision.

Unique Environmental Stresses

The closed environment and limited resources of a space station or space colony introduce special health problems in addition to those already discussed in the context of microgravity, radiation, and human behavior. The task group addresses a few that demand additional investigation. Others can be expected to arise.

Microorganisms

The fungi, bacteria, and viruses carried in flight will experience the unique stresses and opportunities of a small, closed environment. The microbial populations carried in flight by man and animals will be subjected to unknown and highly variable fields of mutagenic radiation. Rotation of crew members on the Space Station will introduce different strains of microorganisms that could contribute to the emergence of new strains of opportunistic pathogens through mutation and genetic exchange.

Experience involving Apollo and Skylab indicates that microbial exchange commonly occurs among crew members. Although inflight infections were neither unusual nor increased during the Apollo missions, several bacterial-associated diseases were experienced by the crew in Skylab 1. In this case, the environment became heavily contaminated with bacteria, reaching 4350 bacterial colony-forming units per cubic meter in the cabin air two days before mission termination. In the course of eight STS flights, numbers of bacteria in the spacecraft air ranged from 200 to 1300 colony forming units per cubic meter of air. *Staphlyococcus aureus* and *Aspergillus* species were commonly isolated from air and surface samples in these flights. No information is available with respect to viruses. We have little information about the survival of microorganisms in droplet nuclei in microgravity. It is likely that microorganisms will not sediment in microgravity as occurs at one-g. This would result in persisting aerosols and high microbial densities in cabin air, especially with inefficient air filtering systems.

Aerosols and Particulates

Aerosols and particulates contribute to the dissemination of microorganisms and of toxic compounds. On the Skylab 3 mission, particulates reached 35,000 per cubic meter on day one in the mid-deck. In the flight deck, the particulate count reached 24,000 per cubic meter. More effective systems to filter air and water and to remove microbial aerosols and particulates must be developed. Increased quality of housekeeping, hygiene, and waste management must be developed.

Relatively few skin-related problems have been encountered thus far in manned spaceflight in spite of the difficult conditions for the maintenance of skin hygiene imposed by the relative lack of water in space. The skin has a high level of cellular turnover, with significant shedding of tiny particulates into the environment. Bathing facilities, personal hygiene procedures, and clothing should not contribute to the particulate load.

Toxic Volatiles

Crew exposure to several categories of toxic substances may be anticipated from several sources: (1) leaks and spills from storage tanks, chemical reagent stoves, and life support/flight control systems; (2) volatile waste products from crew, experimental animals and plants, as well as from nutrients necessary for their support; (3) pyrolytic products derived from excessive heating or combustion, as in small electrical fires; and (4) outgassing of spacecraft materials such as electrical insulation, paints, lubricants, and solvents, and degradation of nonmetallic materials.

The release of any toxic substance into the spacecraft is more serious and hazardous than similar incidents on Earth for a number of reasons. The closed spacecraft atmosphere with its finite volume and low gas exchange rate allows for greater than normal concentrations of toxic materials and thus greater crew exposure with time. Exposure will be continuous, permitting increased accumulation and greater hazard.

NASA's current approach to the spacecraft toxicology problem includes careful initial materials selection as well as outgassing testing for toxic volatiles prior to approving use of these materials in the spacecraft. In addition, inflight "grab" air samples are obtained periodically and returned for postflight analysis. These procedures have proven adequate for Skylab and the shorter Shuttle flights thus far. A large number of volatile chemicals have been detected during flight, mostly within threshold limit values (TLVs) and NASA spacecraft maximum allowable concentration (SMAC) limits. These values, however, are based on modified Occupational

Safety and Health Administration (OSHA) limits for short-term exposures. Although helpful, they should be reexamined in the context of long-term space missions. Much ground-based information must be obtained regarding the long-term, cumulative effect of representative families of toxic chemical products and their subtle effects on behavior and performance.

Water, Air, and Temperature

In future long-term space habitats, including the Space Station, water from Earth will have to be reclaimed and recycled from urine, spent wash water, and habitat humidity condensate. In Skylab, waste water was not reclaimed. Potable water was merely stored, yet it was considered tasteless because it had a reduced level of dissolved gases as well as excess residual iodine required to meet bacteriological standards.

Reasonable standards regarding concentrations of dissolved gas, organic compounds, inorganics, and microorganisms in water must be established. A major goal of the CELSS program is to meet these standards.

In addition to the potability of water is the related question of atmospheric water (humidity) and the key parameter of temperature. The task group has already emphasized that air quality must be controlled in terms of toxic volatiles, aerosols and particulates, and microorganisms. In addition, humidity and temperature are fundamental to human performance and comfort; while extremes of both can be tolerated for a short time, long-term performance and well-being require the definition of a comfort zone and the control of these two parameters. Lack of such control can quickly produce debilitating and even dangerous conditions.

Proposed Research

Although the health maintenance problems discussed in this section are primarily technical, their solutions require an understanding of biology and medicine often lacking in an engineering group. The task group stresses that these problems should be addressed early in the design of spacecraft and in frequent consultation with the biomedical community.

Problems that merit special collaboration include:

1. Screening materials for outgassing and pyrolytic products.
2. Developing a sophisticated onboard atmosphere-monitoring system.
3. Developing air scrubbers, filters, and catalytic detoxicants.
4. Developing decontamination procedures to deal with spills of toxics and of radioactive materials.
5. Designing safe havens and oxygen face masks.
6. Evaluating appropriate medications to inactivate toxic substances within the body.
7. Completing the research required to establish new NASA spacecraft maximum allowable concentration (SMAC) limits in the context of the Space Station and other long-duration missions.

7
International Cooperation in Space Life Sciences

When Congress established NASA in 1958 by enacting the National Aeronautics and Space Act, it provided that space activities should be conducted so as to contribute to cooperation by the United States with other nations. Since then, the United States has signed more than 1000 agreements with more than 100 nations for cooperative space activities. Drawing on the expertise and talents of other nations has often yielded better dividends for each participant—greater rewards than if each nation had undertaken the same project individually.

There are also other subtle advantages. Cooperation and coordination prevent diversion of valuable resources in needless duplication of efforts. Nations that might otherwise have found themselves competitors have worked together, toward common objectives. The task group feels that these advantages outweigh the obvious problems of integrating and coordinating different languages, procedures, and perspectives.

NASA's international programs fall into two major categories: cooperative projects and reimbursable services. The cooperative activities range from flight of foreign-built spacecraft to ground-based study and analysis of data. Also included are contributions of experiments on payloads to be flown by NASA, joint projects to develop flight hardware, analysis of data provided by NASA

Shuttle flights and satellites, training, scientific visits, and joint publication of scientific results. NASA also provides services for which the user country pays; these range from space launch services for which the user country pays; these range from space launch services to data and tracking services.

NASA's international cooperative effort contributes to the U.S. aeronautical and space research program and broadens national objectives by:

- Delivering cost-sharing benefits and complementary space programs.
- Stimulating scientific and technical contributions from abroad.
- Enlarging the potential for the development of state-of-the-art technology and theory.
- Extending ties among scientific and national communities.
- Supporting U.S. foreign relations and foreign policy.
- Minimizing duplication of effort.

International cooperation is reciprocal in that U.S. scientists are also able to fly on foreign spacecraft, as has been the case in the past with American life scientists using Soviet-Cosmos spacecraft.

The life sciences have been prominently featured in international programs, probably more so than any other space science discipline. The following is a list of international life sciences agreements in force or pending.

1. ONGOING AGREEMENTS AND MEMORANDA OF UNDERSTANDING (MOU)

a. CNES (French)/NASA. Life Sciences Working Group meets twice per year (senior managers) to discuss joint special collaborative programs, experiments, and/or hardware.

b. U.S.S.R./NASA. Cosmos flight experiment series.

c. DFVLR (German)/NASA. Same as item a above.

d. Canada/NASA (MOU). Relates to space adaptation syndrome and depth perception experiments and flight of Canadian payload specialists.

e. U.K./NASA. Joint experiments to fly on the International Microgravity Lab (IML) on Spacelab.

f. Australia/NASA. Same as above. Scheduled to fly on SLS-2.

g. Switzerland/NASA. Same as above. Scheduled to fly on SLS-2.

h. ESA/NASA. Joint experiments to fly on IML-1; life sciences and materials processing.

i. DFVLR, ESA/NASA. NASA life sciences experiments on the German D-1 mission.

j. DFVLR, ESA/NASA. NASA life sciences experiments sharing German/ESA experiment hardware on German D-2 mission.

k. NASDA (Japan)/NASA. Spacelab-J (1/88). Japanese use of NASA life sciences hardware, with some NASA life sciences on board.

l. NASDA (Japan)/NASA. Life Sciences Working Group to discuss joint collaborative programs.

2. AGREEMENTS PENDING

m. India/NASA. To fly an Indian payload specialist to conduct Indian life sciences experiments in 1986.

n. Israel/NASA. To fly Hornet Experiment.

3. AGREEMENTS FINISHED

o. Hungary/NASA. Radiation experiment: Shuttle, 1985.

p. France/NASA. French payload specialist; 1985.

q. Italy/NASA. Otolith experiment; early 1970s.

With the advent of the Space Station, with its enormous complexity and expense, international cooperation on a large scale seems advantageous and, indeed, necessary. The management of such a complex system with many users from many nations, all with vested financial as well as scientific interests, presents a significant challenge. International information exchange is needed on what the Space Station is to be and how it can be used. Data on experimental results will need to be exchanged. Experimental priorities, methods, and standards must be established on an international scale, as well as the distribution and use of crew time in orbit. Scientific objectives must be reviewed and priorities set for the use of space, power, time, and common or standard equipment. There is a need for international bodies to oversee and coordinate such an effort if the massive projects now envisioned are to become reality. Examples of such bodies include space agency committees, national academy committees, international committees, independent internal institutes, and existing international organizations (e.g., COSPAR).

8
Instrumentation and Technology

INTRODUCTION

The ultimate success of the research described in the preceding chapters will depend upon (1) the appropriate applications of existing technologies, (2) the development of new ones, (3) a refined understanding of the interactions between humans, computers, and machines, and (4) the optimal reduction, archiving, and retrieval of data.

Many life sciences disciplines (and probably others as well) would benefit from a "return capsule" with capabilities similar to the Soviet "Progress" capsule. Such a craft would have the ability to deliver expendable systems and return samples, specimens, wastes, and experimental materials.

For clarity, the task group addresses each of the five disciplines—exobiology, global biology, space biology, space medicine, and CELSS—separately. *However, many of the needs, especially those of computation and robotics, are common to more than one discipline.*

EXOBIOLOGY

General

The requirements for special instrumentation and technologies for exobiological investigations cover a very wide spectrum, from unique ground-based laboratory facilities and techniques to dedicated space missions to other objects in the solar system.

Specific Needs

- Microchemical techniques for the identification of materials in individual microfossils.
- Highly sensitive mass spectrometric techniques for the identification of compounds and isotopes.
- RNA synthesizers, similar to those already available for the synthesis of DNA.
- Laboratory simulators: Several kinds of simulators would be useful for trying to understand the course of chemical evolution. These range from devices to suspend one or more particles (to simulate interstellar grains), to large simulators to study the formation and evolution of organic compounds over long periods of time.
- Collectors for cosmic dust particles: methods need to be developed for use in space that would provide for the nondestructive capture of interstellar and interplanetary particles.
- Rover technology with special emphasis on optimizing interactions with remote human operators.
- Technologies for the collection and subsequent handling of extraterrestrial samples.
- Hubble Space Telescope (HST): This new tool could yield substantial information for exobiology. It could be used to inventory biogenic elements in external galaxies, to image Titan in the ultraviolet and visible range, to search for organic molecules in Titan's atmosphere in the ultraviolet, and to study tenuous atmospheres in the outer solar system.
- Space Infrared Telescope Facility (SIRTF): This instrument could also provide an enormous amount of new data for the exobiologists, especially if its spectral and spatial resolutions could be improved over those currently being planned. SIRTF could be used to identify and quantify the molecular constituents of the

jovian planets and their satellites and to study the composition of intersteller grains, comets, asteroids, and external galaxies.

- Large Deployable Reflector (LDR): This instrument could give information in the far infrared and millimeter regions of the spectrum. Its utility would be for study of protostellar objects, novae and supernovae, and the interstellar medium.

GLOBAL BIOLOGY

General

Remote sensing is essential for research in global biology. In order for the results from sensitive spectrometers and other sensors to be interpreted accurately, they must be standardized against ground-based measurements. Sensing capacities must be developed to determine the biome distribution and productivity over the entire surface of our planet. Passive reflectance spectrophotometry in the ultraviolet, visible, and infrared, can supply much of this information.

Specific Needs

- Spectrometers in the visible and near-infrared with high spectral and spatial resolution.
- Color imagers with high spatial resolution.
- Laser fluorescence sensors for use in aircraft and spacecraft.
- A synthetic aperture radar device for spacecraft studies of surface water and plant structure.
- "Close probes" or penetrators: Instruments designed to make specific global measurements at remote or otherwise inaccessible sites on Earth.
- Polarization photometers.

SPACE BIOLOGY

General

Experiments in space biology and in space medicine will be expensive; opportunities will be limited. They must be preceded by thorough evaluation of ground-based model studies, for example by the clinostat or by bed rest. When eventually performed at microgravity, these experiments will require one-g controls in order to be reliably interpreted. Whenever possible, humans should be the subjects for the experiments for reasons both of economy and relevance. Even so, holding facilities specifically designed for a few species of plants, animals, and microorganisms must be provided.

Specific Needs

Solution of the interrelated problems of phototropism and geotropism requires the development of sophisticated plant growth chambers similar to those of CELSS. A range of variables must be monitored and, in some instances, regulated. These include: light intensity; period and spectrum; oxygen, water, carbon dioxide, and ethylene concentrations; and evolution of trace gases, some of which are not yet identified. Similar gas monitors are required to monitor the air of cabins to detect leaks or outgassing. They are also invaluable as research tools—applied to breath, sweat, urine, and feces of animals and humans—to monitor physiological state and general health. Provision must be made for freezing or lyophilizing specimens as appropriate for return to Earth for definitive analysis.

Facilities to house experimental animals, including primates, for long periods of time on spacecraft will also be necessary. These facilities should be designed so as to provide structural isolation from humans but to permit functional interactions when necessary for animal care and for experimental purposes.

For many investigations, including growth of crystals of proteins and of nucleic acids, it will be necessary to conduct experiments in an environment with approximately 10^{-6} g and <1 Hz. Such a locus is theoretically achievable at the center of the currently conceived Space Station. However, with projected growth (and activity) on the station, changes in the center of mass could pose significant problems. For these reasons, consideration should be given to the design of free flyers to provide the required low-g environment.

Sensorimotor experiments, especially as they focus on the complex integration of sensory information from the inner ear, the eyes, and proprioceptors throughout the body, require access to a centrifuge and a linear accelerator that will accommodate humans and primates. These experiments also require a range of electrophysiological equipment, including the placement of semipermanent electrodes in the brains of rodents and primates.

In order to distinguish the effects of linear acceleration as it affects the otolith, and of rotation as it affects the semicircular canals, it is also essential to have a linear accelerator large enough to seat human subjects.

Several additional types of variable-force centrifuges will be necessary to conduct life sciences research in space. The centrifuge to be used for vestibular/neurosensory research might serve as an operational device to prevent and treat spaceflight "deconditioning" in humans. For these, it may be necessary to consider a long-radius tether design. For centrifugation of smaller animals, a short-radius machine may suffice. In addition, consideration should be given to the design of a large-capacity, continuously running centrifuge at (or near) one-g to maintain plants and animals until they are chosen for weightlessness studies. Most importantly, a centrifuge designed to run inflight controls is essential for reliable interpretation of both plant and animal experiments.

Prosaic as it may seem, the task group emphasizes the importance of binocular low-power and high-power microscopes for observing tissues and cells. These should have dark-field and fluorescence capability and be equipped with video for documentation and for direct transmission to sponsoring laboratories on Earth. Quality microscopy demands quality sample preparation—preservation, sectioning, and staining—as well as maintaining living cells.

SPACE MEDICINE

General

Most of the proposed experiments, as well as monitoring of astronaut welfare, rely heavily on noninvasive and microchemical techniques. They also require an extensive data base to define "normalcy" at one-g and "normalcy" at zero-g. Consistent with her or his privacy, dignity, and function, every crew member should, as a condition of accepting the assignment, participate in establishing this data base. Their participation is much less expensive than maintaining an extensive colony of primates or than failing to detect the first signs of illness.

The psychology of group interactions will contribute to the assignment of crew, design of living and work spaces, and scheduling of activities. The unique circumstances of extended spaceflight will also contribute to the study of behavior. Again, consistent with established guidelines of medical ethics, the documentation and monitoring of behavior should be anticipated. Innovations might include correlation of various physiological and metabolic measurements and correlation of these, in turn, with composite indices such as activity, frequency of talking, and voice inflections.

Specific Needs

Noninvasive imaging techniques are required for research in all areas and will be valuable for health care. Echocardiography has been used in Shuttle flights to assess cardiac chamber size and function. Future ultrasound developments will include improved resolution, incorporation of function into the anatomical picture, and perhaps even the ability to measure not only internal anatomy and flow, but also, indirectly, pressure. Ultrasound will be important in bone work and in muscle studies. Newer ultrasound techniques may be developed to image discrete muscle groups and correlate images with muscle tone and elasticity. Lightweight photon spectrometers can improve the scope of bone research. New methods that merit exploration include: electron spin resonance (possible after ingestion of bone-specific dyes containing free radicals) and piezoelectricity (of teeth). Computerized axial tomographic (CAT) scanning could have great utility in bone and muscle research if smaller, lightweight instruments are developed. The same is true for nuclear magnetic resonance (NMR) techniques, which have the additional virtue of being able to monitor biochemical changes in cells.

These imaging techniques should be complemented by the development of several minimally invasive or noninvasive physical monitors and chemical techniques.

Important measurements include temperature, heart rate, blood pressure, oxygen tension, pH, electrolytes, osmolality of perspiration, and muscle tone. The possibility of wearing a simple "smart band-aid" or wristwatch that could measure and even process some of these data from skin should be pursued.

Microchemical analyses should be developed for urine, saliva, sweat, and small volumes (0.01 ml) of blood. Measurements should be made of standard metabolites, byproducts of drugs, various hormones, antibody titers, and enzymes. Various of these reactions can be performed "dry," that is, with reagent-impregnated papers and with extremely sensitive and specific antibody-coupled reagents. These physical and chemical procedures will serve onboard health care as well as primary research functions.

Recent and projected advances in immunochemistry and cell culture should be utilized. The task group anticipates the need for a miniaturized laser-cytofluorograph for the analysis of lymphocytes and other cell types. Automated instrumentation incorporating monoclonal antibodies and antibiotic sensitivity should be developed for microbial identification.

Several general monitoring systems are required as integral components of health maintenance and safety. Cabin air should be routinely monitored for particulates, aerosols, bacterial count, and various trace gases. Radiation, especially heavy ion particles, must be measured. A protected multipurpose area to be used to counter radiation and toxicological exposures, fires, and loss of cabin pressure and for quarantine of communicable disease should be designed.

This combination of analyses and monitors is required both for research and for improvement of astronaut safety, health, and performance. These data must be integrated to help understand the whole person; the final monitor is behavior and performance. The task group recommends the development of objective and predictive monitors of behavior such as frequency and type of activity, and voice pattern analyses.

CONTROLLED ECOLOGICAL LIFE SUPPORT SYSTEM (CELSS)

General

There is considerable overlap between the technology and instrumentation required for space- and ground-based research in CELSS, and that required for space biology and space medicine. Therefore, the task group focuses here on those facilities that need to be developed.

Specific Needs

Essential to the development of a Controlled Life Support System is a growth chamber for higher plants (a "phytotron" or "biotron"). The building of such a facility, with its control systems for lighting, gas sensing, and nutrient and water delivery in space, would be a major undertaking. Since it is unlikely that this system will also satisfy the needs of plant researchers interested in gravitational biology, additional plant growth chambers will need to be developed.

The task group cannot yet specify a controlled ecological system that would be adequate to supply portions of the astronauts' food, water, and oxygen, and to recycle their wastes. Many ground-based studies must be initiated now and completed well in advance of even preliminary designs for a space vehicle that could recycle a significant portion of its water and organic compounds. As noted earlier, we cannot yet choose the most appropriate plants, algae, or yeast, nor make reliable estimates of the quantity and quality of their edible products.

These plants and cultures of algae or of genetically engineered cells are continuously operating chemical factories and require automatic controls and sophisticated external monitors. These include: measurement of pH, conductivity, spectra (absorbance and fluorescence), dissolved molecules such as urea, glucose, and plant hormones, and evolved gases such as water vapor; oxygen, and ethylene. Human observation of higher plants may be the best monitor. An examination of individual cells or squashes for karyotype changes and chromosome staining patterns will be required to monitor mutations and transpositions, especially in artificially created cells.

The optimal use of these cultures and higher plants requires not only a better insight into the photosynthetic organisms themselves, but also a refined understanding of human nutrition under flight conditions and of the food technology that will process and supplement the products of photosynthesis. Methods need to be developed for the harvesting, milling, mixing, and preparation of foodstuffs. Fluid handling technologies need to be developed for use in a weightless environment. This applies both to liquids (e.g., how to best deliver water to plants) and to gases (e.g., how to separate and store biologically produced oxygen and carbon dioxide). Control system methodology must be designed for CELSS that will maintain an essentially constant environment even though the elements of the system have different time constants for generation and uptake. Waste disposal and waste utilization techniques for the conversion of inedible plant products into recyclable materials need to be developed.

COMPUTATION, INTEGRATION, AND ROBOTICS

The experiments described, especially remote sensing for global biology, will generate up to 10^9 bits of data per day. It is imperative that the appropriate strategies for formats, data reduction, onboard computations, data transmission, and data archiving be determined well in advance. In general, precedents of other disciplines should be honored. Biological experiments and health maintenance pose several unique problems. Much of the work is interactive in that a subsequent step cannot be decided on until an initial result has been analyzed and discussed with an earth-based laboratory or clinic. In particular, microscope images or physiological monitors may have to be transmitted in real time, anticipating a prompt response. This may require interactive graphics with mutual manipulation of "drawn" objects. Organisms—humans included—are complex; the evaluation of experiments on them requires the integration of much information.

Humans, ground-based, or as crew members, must interact with instruments and pieces of equipment that contain complex mechanical components and powerful computers executing complex algorithms. Although in principle this situation is hardly novel, the extreme conditions of space research and construction dictate a major investment now in robotics. It is inappropriate for this task group to address the political, emotional, or ethical aspects of manned spaceflight. Nevertheless, any rational analysis of man's role in space must include an analysis of robotics. The issue is not man versus machine. Inevitably, both are involved; this task group seeks to optimize their interactions. A simple mechanical hand can grasp with strength under extremes of temperature and radiation. The human eye and brain can recognize instantly an unanticipated signal hidden in a noisy image. How best to transmit tactile information from a distant "hand?" How to correlate these with video images? What response times can be tolerated—seconds or days—in the sensation or the command sequence? In addition, work is needed in the area of teleoperations—the creation of a real-time interaction by radio/TV link between the ground-based investigator and the on-board experiment or surrogate experimenter. This capability will allow experiment revision or modification in much the same fashion as on the Earth. The answers to these questions should have profound effects on the future of all the space sciences.